Mister Blister

John Rayment

May the sun come out
to greet you and
the pubs not to
far apart

John
Rayment

ATHENA PRESS
LONDON

MISTER BLISTER

ISBN 1 84401 073 2

First Published 2003 by
ATHENA PRESS
Queen's House, 2 Holly Road
Twickenham TW1 4EG
United Kingdom

Printed for Athena Press

Mister Blister

Desiccation

Stretched round the Cornish coast o'er dale and hill
The South West Coastal Path means no man ill
Vast sandy beaches children's dreams to fill
While surfers ride the breakers' foaming mill
And Jack heads for the sand dunes with his Jill
We're young (ish), healthy, fit and full of skill
No need to plan, surprise gives greater thrill
But constant ups and downs sap strength and will
The sea wind's breath blows lonely, cold and shrill
A bitter lesson learnt but with us still
The slightest blister makes me fit to kill
To walk on one, the pain would grow until,
Unless some vital mission to fulfil,
I'd sit right down and say 'I'll wait here 'til
You fetch me water, bandage and a pill!'
What then, if broken blisters, raw and drill
Each shred of skin, and yet more form a sill
Blood, sweat and gore seep from each ridge and rill
While your mate laughs and mocks with constant trill:
'Now, stop that fuss, you blister-ridden dill'
To tell this tale is why I lift my quill
So come with us, you'll find the journey brill
(OK, the poems are not fit for swill)
I'll write the words, but who should head the bill?
Mister Blister, my walk companion, Phil

Contents

Well Chuffed

As the train pulled in to Reading station, I searched the platform for Phil, but could only see ordinary looking people...and what looked like a pile of old rags someone had left for a jumble sale.

Suddenly half the pile stood up and became recognisable as a human being, wearing a pair of worn down walking boots, caked in mud and muck from some previous trip, and crumpled blue walking trousers, with numerous holes of various shapes and sizes, and a vast number of ageless, unidentifiable but worrying stains. Similar care had been taken of his screwed and stained dark blue T-shirt, with London School of Economics (LSE) logo.

Wrapped round his waist were a red fleece, which looked as though it was the genuine article, with half the sheep still attached, and a large blue bum bag, worn at the front, stuffed with all sorts of odds and sods, fit to burst at any moment, with a safety pin holding the broken zip together. An LSE baseball cap, peak at the front, and surprisingly normal, boring glasses, completed his attire.

Phil turned to the rest of the pile, which was actually his rucksack, with pieces of clothing and equipment trailing out of every pocket, and picked it up by the back straps. Whereupon, unnoticed by Phil, a walker's compass and half-eaten toffee bar fell to the floor. They were retrieved by another passenger who gingerly handed them back, his body language clearly showing he wished he had fought his natural friendly, helpful instinct, and was in fear of catching some dread disease.

Phil threw the rucksack onto the train, placed the compass in a trouser pocket and the toffee in his mouth, having wiped it 'clean' down his disgusting trousers, and climbed on board. We were setting off on a three-week walking holiday in Cornwall, starting at Newquay, on the Atlantic coast. Somehow I had allowed myself to be persuaded by Phil that it would be a fantastic start to go there via a steam trip on *The King* to Kingswear, which is in

Devon and on the English Channel coast.

I love an early morning start to an adventure, so when my watch alarm signalled 5.00a.m. that morning, I had leapt out of bed, dressed, struggled into my rucksack, kissed Lynn goodbye and set off for Gidea Park station. I could have requested a lift, and am sure Lynn would have been delighted – to be given the chance to tell everyone how I had started a three-week walking holiday by cadging a lift to the station.

It was a beautiful, blue-sky morning, with the sun already pleasantly warm and the streets quiet and deserted – I saw only one other person, walking his dog, two cars, and a fox. The latter came confidently out of some bushes edging a park, and proceeded to trot along just a few yards in front of me, as if showing the route or keen to come along, before disappearing through a gap in a fence onto some rough ground. I took this friendly welcome from a lucky fox as a good omen – given my normal lack of planning; I knew I would need plenty of good luck. My rucksack was heavy, but certainly no worse than I had expected, and gave me that self-righteous and proud feeling shared by people carrying their own gear on their back. I felt good. Foxy.

I had decided to wear my walking sandals for this easy day, saving my boots for the real thing. By the time I got to the station one of my feet felt sore, but I had a much bigger problem to deal with: the station was locked.

I was ten minutes early for the train, so initially assumed someone would appear to unlock in time for me to stroll down onto the platform and get on board. After a couple of minutes a middle-aged workman in a donkey jacket appeared:

'He's asleep again, is he? Come on, we'll have to climb in.'

We walked round to the back entrance of the station, where the regular climbed over an eight foot wire fence and dropped onto some waste ground, then onto a six foot wall so as to squeeze beneath another wire fence and drop down onto the walkway leading to the platforms. I somehow managed to pull / push / shove / levitate my self and rucksack over these little testers, wondering whether the fox had indeed been a lucky omen, but eventually persuading myself that I had been lucky: the door had been locked, but my donkey jacketed hero had saved the day, and fighting my way through an obstacle course had left me feeling good, and pleased with myself.

By the time I reached the platform, Mr DJ had woken the stationmaster with a few sweet nothings whispered in his ear. Yawning and scratching his head, he lazily strolled off to get his keys just as the train was pulling in. I watched him slowly lumber up the steps to unlock the door, whereupon a bunch of not amused passengers piled down the steps shouting friendly greetings to the driver as their train started to pull away.

After about ten yards, the train stopped and the doors opened to let everyone on, by which time the stationmaster had returned and was waving us all on our way with a big grin on his face. I have no way of knowing, but I had a strong suspicion, that this was all a routine put on to entertain the customer, and the driver had no intention of leaving his work mate to face the mob. I do know it brought a smile to my face, which increased when I looked out the window to see, sitting halfway up the grassy bank, my fox, head held high, watching the train pull away.

Obviously I had not had a chance to buy a ticket, so when we arrived at Liverpool Street and the barriers were open with no staff around, I left my fare on the shelf outside the ticket booth, honest. I had forty minutes to cover the eight underground stations to Paddington, where I was to catch the *Kingswear Special* steam excursion train, Phil joining at Reading.

At the underground entrance I discovered my next surprise: the underground doesn't start until 7.00a.m. on a Sunday. Still, no problem: catch a taxi. I found the rank, but no taxis, so wandered around the deserted streets, with no real plan, pondering all the sound reasons why any sensible taxi driver would be in bed, and wondering what my chances were of getting to Paddington in time by bus, assuming there were any at that time of day.

After what seemed like ages, I was beginning to get desperate and lose faith in my lucky fox, when a taxi came round the corner – I made Paddington with fifteen minutes to spare: excellent planning had ensured I was in good time for the wonderful steam trip.

The *Kingswear Special* was listed on the departures board, but without a platform number. I tried not to make it obvious I was waiting for the steam special in case someone recognised me, but at the same time I tried to pick out the steamies from the normal people, which proved to be quite easy. At 6.54a.m., six minutes prior to scheduled departure, the platform number appeared on the board, and I set off, followed by a train of steamies, to search for the relevant platform.

Signs seemed to point to every platform but the one I wanted. Eventually I discovered it on the far side of some major redevelopment work, with scaffolding and tarpaulins hiding the signs and preventing access through the obvious route, necessitating an improvised detour. By now at least twenty steamies were following me, and I prayed I found a route through in time for them to catch the train, not wanting to find out how they would react if they missed The King.

My seat was in coach D. I went past A, B, C, E. What is happening? If all else fails, ask someone in a uniform:

'Coach D? Yeah, mate. That's the Past Time Members' club coach. It's up the front.'

Later, it transpired that Phil was a member of the Past Time Members Club and had obtained seats for us in their special 'olde worlde' coach. Their president had had it moved to the front of the train so club members would be travelling as close to the engine as possible, receiving maximum benefit from the noise, smoke and grime. We were in front of the Premier and First Class coaches, in which the idle rich were being served champagne breakfasts, to be followed later by a four course dinner.

I soon found out that part of the deal was club members were not to go through those carriages in case they disturbed anyone, or interfered with the waitresses. What kind of people were these Past Timers, if such restriction had to be placed on their movements? Did their club name really allude to the fact that they were all ex-criminals, having done time in their past? Or was the restriction made to emphasise the class split, similar to the way a curtain is always pulled across between cattle and club on a plane, which also helps maintain the mystery and myth of exactly what goes on in the upper classes: heaven forbid the masses realise they actually receive virtually the same service for a fraction of the cost – they already know they arrive at the same time.

Whatever the reason, it meant we could only gain access to the buffet car when the train stopped at a station (as opposed to, as I was to discover, all the other stops it made for no apparent reason). Of course, no one told me this until the train was moving, so I could not even get a cup of coffee. Normally, I doubt if I would even think of having a coffee for the first hour of a trip, but now I knew it was available in the buffet car, and I wasn't allowed through the first class carriages to get at it, I suddenly felt very thirsty and downtrodden.

I thought of complaining, but all the other punters were extremely happy with the arrangement: who wants to be distracted by such worldly things as food and drink when they are travelling behind steam? I also noticed that many of them were wearing red jumpers with the rather worrying logo 'Past Time Political Wing' emblazoned on their chests, proudly referring to themselves as PW. I did not want to be the morning's sacrifice to the Iron God, so decided I wouldn't say anything until I had support when Phil joined at Reading.

I also decided to wait until I could ask Phil to explain the other thing that was bothering me – the fact that we were being pulled by a diesel! (Later, Phil informed me it was a class 47). I had obviously missed something, but weren't we supposed to be on a steam trip? This seemed too obvious a point to have escaped everyone else's attention, and I knew I faced derision, sarcasm and wit if I started asking dumb questions of these dedicated soldiers of steam.

Initially, the carriage being empty, I had a section of six seats and a table all to myself, but when the carriage filled up, two men came to sit opposite me. Both looked relatively normal, apart from badges sewn into their jacket lapels and hats advertising previous trips, particular branch lines, and famous locations. They were friendly enough, attempted to strike up a conversation with me, but it reminded me of my 'O' level French oral exam, when another stranger had sat across a similar-sized table from me and spoken a jumble of sounds which I had failed to sort into individual words, let alone meaning. In some respects this was even worse, in that they were using my language, but I was unable to understand much of what they said. Shunting bogies, letting off, two-four-two, double heading, slipping flies…?

I started to explain that I wasn't really into trains, but was meeting a friend at Reading who was, and we were off on a three-week walking holiday. It didn't sound convincing, even to me, so I was not surprised to receive a series of looks which conveyed they thought I must be some kind of a nutter, not worth wasting time on. Very quickly, they chatted to each other, while I was left on the other side of the table, a social outcast. Being treated with contempt, mixed with a little pity, by a couple of steamies was an unnerving experience, but when I looked around for someone else to chat with, I realised these two were easily the most normal-looking people there.

I was glad to be out of the conversation, as I still feared making some horrendous gaff. To fill the time, I started to wonder what other badges these two might have on their jumpers, and soon had a mental image of Hell's Angel style 'Steamie of Death' tattoos across their backs. Unfortunately, once my imagination sets off, it can get out of hand, so it was not long before they were festooned in a net of criss-crossing railway track chains connecting various forms of steam related body piercings attached to their ears, tongues, nipples. Had they been a couple of beautiful women, this may not have been too disturbing an image, but as it was, it was rather repulsive and disconcertingly hard to shake off.

I was pleased to find Reading station approaching, along with the apparent pile of old rags, which had now picked itself up and was making its way to join me in the Past Time club coach. I had begun to fear that this was all some kind of set up – Reading station would come and go with no Phil, leaving me to the mercies of the Political Wing! Phil is not the type who would do that kind of thing, but I certainly am, as are many of our mutual friends, so I guess I had been scared some of them might have knobbled Phil. If I had been thinking rationally, I would have known there was no way Phil would give up a steam train trip so easily.

As he made his way down the corridor, Phil managed to only bash three people with his rucksack; none objected, presumably because they all seemed to know him from past trips:

'How are you old thing?'

'Haven't seen you since York...'

'Do you know that odd looking bloke in the corner?'

'Hear about Scot, poor bugger?'

'Clocked up six thousand miles since May...'

'He's with me.'

'Where's your PW jumper?'

'PW, PW, take care we don't trouble you...'

'Well, he doesn't look like one of us.'

'In my bag.'

'Did you make Bristol?'

'Can I stuff him in the boiler?'

'Why all the kit?'

'He's all right. We're going walking in Cornwall.'

I assumed they were joking and wouldn't really stuff me in the boiler, but something else did concern me. Suddenly I felt as though I might be the odd one, and these steamies normal! Once the greetings had finished, we returned to my table, and Phil proceeded to spend half an hour deep in foreign conversation with the two men opposite.

At the next stop, we transferred to the buffet car where I bought myself a long awaited coffee and sandwich, but then had to stay there until Didcot, where *The King* at last put in an appearance, and we were robbed of our strong, solid and reliable proper diesel engine. Don't tell the Political Wing I said that.

Phil attempted to get me up to speed, while I pretended to stifle yawns and made sarcastic interruptions, although actually being quite interested:

'The train is *The Torbay Castle*.'

'Really?'

'The engine being known as *The King*...'

'Fascinating.'

'But her official name is *King Edward the First*, and she is engine number 6024.'

'Wow, how interesting.'

'You are privileged to be part of this special occasion.'

'You think so?'

'*The King* rarely goes down the branch line from Paignton to Kingswear…'

'Incredible.'

'Which is owned by the Paignton and Dartmouth Heritage Railway…'

'Snore…'

'And is the final part of today's trip.'

'I can't wait.'

'It will take eight hours to get to Kingswear, so we have a full day's steaming ahead of us.'

'Is that a threat or promise?'

Suddenly, something Phil had just said registered:

'Wait a minute. Did you say her name is *King Edward the First*, and she is engine number 6024?'

'Yes'

'*Her* name…'

'Yes,'

'*King Edward the First*?'

'Yes…'

'So *The King* is a she?'

'Yes. All steam trains are female. Same as ships.'

'What, even male ones?'

'*The King* is not a male train.'

'I know he is not a mail train – letters have to be delivered on time!'

'Very funny. She is called *The King*, but is still a woman.'

'Sounds a bit odd to me.'

'Well, it isn't'.

'Bit of a trainsvestite?'

'Tee hee, how original. You'll be asking if she's gone off the rails, lost her driving rod, taken a turn for the worse, or been shunted up the rear soon. Don't you think we've heard them all before? It's part of the romance – all ships and trains are women – some men say because they are temperamental, take ages to get ready, and have half a mind of their own, but I believe it is because they are incredibly graceful, beautiful and you can't help falling in love with them.'

Certificate No. 008 No. of Shares.....One.

SHARE CERTIFICATE

6024 PRESERVATION SOCIETY LIMITED

Registered under the Industrial and Provident Societies Act 1965

Registered No. 21746R

This is to Certify that Mr.G.J.Webster

of 3 Woodham Drive, Hatfield Peverel, Chelmsford.

is the Registered Holder of ....1.... *fully paid shares of*

One Pound each numbered 1008 *in the 6024*

Preservation Society Ltd., subject to the Rules, Regulations

and Constitution of the Society.

Signed this 5th day of June 1997

.... Chairman

.... Secretary

.... Treasurer

A large crowd of enthusiasts watched *The King*'s arrival and hitching up, and for the rest of the trip there were people waving and taking photographs from every bridge and station as we passed. I couldn't help feeling like a star and beginning to enjoy the trip. Very worrying, mustn't let Phil know.

Steamies seemed to come in a wide variety of types and specialities. People all round me were discussing which lines they had been on, which engines, speeds up particular inclines, number of water stops... Others were leaning out of the windows with their heads in the clouds...of smoke and steam from the boiler, most wearing goggles to keep out the coal particles which could badly damage one's eyesight. One character looked like he had just finished major repairs on the engine, with filthy black overalls and jacket, a grimy balaclava, and old fashioned motoring goggles, leaving just a couple of patches of skin exposed. He spent most of the journey leaning far out of the window, a microphone in one hand and tape recorder in the other, recording every whistle and roar, to brighten his winter evenings. The image of Toad from *The Wind in the Willows* was swift to mind, and I had visions of him wistfully murmuring 'poop, poop'.

Gradually, a black layer was forming on the tables, our clothes, and every other surface, and pieces of grit were finding their way into coffee cups and sandwiches. The taste of coal was in the air, and in my lungs, and when we passed through a tunnel, the whole carriage filled with a rich sooty smog. Despite, or perhaps because of, all this, I was having a good time. The throb of the engine, with its feeling of raw, untamed power, and exhilarating glorious freedom of the frequent whistles and hoots, were completely spell-binding: this was going to be a great trip after all.

Everyone in the carriage was steam-crazy but harmless so far. Probably the most scary person was a guy who seemed to just sit and stare straight forward, a nervous grin playing at the corners of his mouth, incredibly skinny but constantly winding sandwiches into his mouth like a machine. I didn't dare ask him what his speciality was.

The man sitting opposite me started to record the time as we passed each milepost, then mentally calculate the speed for that mile and write it in a little black book. Two others had a special steamies satellite positioning gizmo, which told them the distance travelled, speed over the last mile (or any other distance) and average speed between any two points, all to several decimal places. It seemed a bit odd to be using ultra sophisticated technical gadgetry when on a steam nostalgia trip, so I decided they were posers and sided with the DIY guy when they started to get a bit superior with the quantity and quality of their information.

I suddenly caught sight of myself in my mind's mirror. Oh my God! I'm not only enjoying the trip, but am now defending a steamy for not using modern technology! I went and sat in the corner to give myself a serious talking to, but it became a frightening experience, as I found I was in danger of scoring minimum in one of those classic jokes one tends to receive by e-mail:

Attributes of Employees

Board Member	Tells God what to do
Top Management	Thinks he is God
Upper management	Talks to God
Senior Management	Talks to others
Middle Management	Talks to himself
Lower Management	Argues with himself
Junior Management	Loses those arguments

During the journey we had to make three stops to refill the boiler. One was at Taunton where the old water storage towers had been removed many years before, and it was now necessary to lay hose from a distant supply along the full length of the platform to the front of the train. I couldn't help but be impressed by the speed and smoothness with which the enthusiasts achieved this, and the fact that the hoses all just clicked together with leak-proof joints, which is more than I have managed to achieve on my shower tray at home, despite hours of effort and vast quantities of plastic trim and goo.

At one point, we stopped for no obvious reason, and the announcement was made that we had just run out of steam, and would have to wait for ten minutes while *The King* built up a head again. At Yelton there was a further announcement, this time that someone had threatened to jump in front of *The King*. I asked Phil if he thought it was a suffragette, but he didn't seem to pick up on the Past Times link and replied 'No, they haven't suffered yet, they only threatened to jump'. We couldn't go on until the police had checked the line, and by the time they had done this, we had lost our 'path' so would have to wait until some (real) trains had gone through. Don't they realise how important *The King* is?

Err, oops!

I noticed several steamies seemed to be getting nervous, and I eventually discovered the cause: we were approaching a very long 1:80 incline, known as Whiteball or Wellington Bank. I thought they must be joking, as I found it hard to believe such a gradual incline could prove problematical, but soon after we started up *The King* began to struggle. Gradually speed decreased, and the puffs and chugs became slower and more laboured. There was a deathly hush in the carriage, as everyone could feel the effort being put in and was silently willing success: I guess we were all saying to ourselves *I think I can; I think I can...* in time with the beats of the engine.

'I...think...I...can;I...think...I...can;...I...think...oooff!'

Silence.

Shit, I can't I thought, but was too scared of the Political Wing to crack such a joke.

No steam. *The King* had run out of puff again, but this time halfway up a long climb – I knew the feeling. This time we waited fifteen minutes while s/he built his/her strength up, then there were a few extremely tense moments as s/he made several attempts to get moving, with each pull shuddering down through the carriages, then echoing back to the engine. Slowly, the forward shudders joined together until a constant momentum was achieved. Having beaten the inertia, we rapidly gathered pace, reached the top of the incline and started down the other side with a long, triumphant whistle. Everyone, including me, let out a great cheer.

Oh my God, let me off quick or I will be signing up for the PW!

All the dramas had combined to leave us sixty minutes late on arrival at Kingswear. *The King* went off for a well-deserved drink and sleep, being replaced for the return trip by one of those new-fangled diesels. Instead of having seventy-five minutes for a quick visit to Dartmouth on the Lower Ferry, we barely had time to look across the Dart. Who cares? We had been behind steam.

The diesel pulled smoothly, cleanly and easily away, then strolled up the dreaded incline without breaking into a sweat: luxury. Steam may be okay for an occasional nostalgic trip, and it is impossible not to start thinking of the engine as a person – male or female – but imagine the chaos if it returned for main line duty. I'm safe: you can stick your Political Wing up your boiler and whistle for it.

The stretch between Exeter and Kingswear had been by far the best part of the journey, hugging the shoreline round a good mix of coves, headlands, sandy bays and estuaries. It had been a gloriously sunny day, and everywhere had been full of happy people enjoying themselves. Our plan was to stay on the excursion train while it retraced that part of the journey, then catch a train from Exeter to St Austell, and a taxi across country to Newquay.

There were several looks of amazement at the size of our packs when we prepared to get off, and several standard 'rather a lot for a day out' jokes ensued. We had an hour and a half to kill at Exeter (St. David's) station, the diesel having arrived on time, but Phil knew every pub in the *CAMRA Good Beer Guide*, and soon steered me to The Great Western Hotel:

> *Built to serve Brunel's St David's Station. The bar is a Mecca for real ale enthusiasts and railway buffs alike. There will be Sunday papers piled on a central table, and several large dishes of nuts and other snacks on the bar.*

Needless to say, these prophecies were fulfilled, but one of those strange, inexplicable things happened: having seen the bar snacks begging to be scoffed, and despite having not eaten much all day, I forgot all about them and just read the papers and supped the beer. I guess my brain was still reeling from the steamy seduction.

On the drive from St Austell to Newquay, only twelve miles but seemed more like one hundred and twenty, the taxi driver took the opportunity to tell us his life story. His wife had kicked him out – she didn't understand him, wanted the money but didn't like the long hours. This was a new taxi firm, much better than all the others; this one had no problem with brake failure. His wife had kept the kids and now expected him to give her nearly all his money, which he did, but it wasn't really fair…

I was trapped in the front, so had to make the traditional 'yes, no, oh dear, really, *whatdoyoumeanbrakefailure*' kind of comments, while Phil pretended to be asleep. By the time we reached Newquay it was 9.00p.m. and we still had nowhere to stay. The driver agreed to assist our search;

'I'll take you to the camp site where I used to work, they'll have plenty of spaces and it's really good.'

We arrived at the gates, to be met by the reaction 'Sorry, no groups of all males.'

Now I know some places are concerned about groups of drunken men fighting or wrecking the place, but you only have to take one look at us to see – well, okay, I take the point. Our loyal driver was still confident;

'There's another site just up the road, not as good quality, but will do for you.'

I shot him a glance

'I mean, given your urgent need.'

Hmm. I enquired at reception,

'Yes, of course. Minimum six nights'.

Presumably this was a special rule for the eclipse period, or maybe applied for all peak periods. I toyed with pointing out the fact that, if they had space at that time of night, they were very unlikely to let it to anyone else. Letting us stay for one night would not block any sites for future nights, so would be all clear profit. I could easily have slipped into one of my favourite lectures: relevant costing. I weighed up the receptionist, decided she was not open to such persuasion, and returned to the car.

Our fairly loyal driver was still quite confident:

'There are no other campsites around, but plenty of bed and breakfasts.'

He drove us back through town, to the top of a hill, round a roundabout, pulled into the kerb, and said, rather pointedly:

'Of course, there is an eclipse next week. Didn't you think of booking ahead? Bed and breakfasts all along here; good luck; goodbye.'

Good grief. It was half past nine and getting dark. It would be nice for at least our first night to be spent in relative comfort and hassle free. At the very first B&B we tried, we received a warm, friendly reception:

'Hello! You two look knackered. Bet you'll be pleased to get in your room, freshen up, have a cup of tea and maybe a quick pint or two in the pub?'

What a relief. I'd just started to think we would be camping on the beach among the lovers, loafers, drunks, surfers and tin cans.

'Certainly will. Shall we leave our bags here while you show us the room?'

'Oh, sorry, we're full. Did you think we had vacancies?'

At the next B&B I was more direct, and at least the rejection was clear and fast.

It was now very dark, and as I walked up the drive to the third, I recalled my lucky fox: now was the time for him to show his metal.

'We have one twin available, but for tonight only. You'll have to move on tomorrow.' Foxy Phew.

We were shown to an excellent room by a very friendly couple. There did not seem to be anyone else around, so we assumed they must all be up the pub, which is where we should be. By the time we had dumped our bags and hit the front door it was 10.00p.m.

On the two-mile walk into the centre of Newquay, I told Phil how it was my lucky fox that had found us the B&B, but he had his own view,

'You always seem to be jammy with travel and accommodation, but I believe in imperfect guardian angels. Yours is great at ensuring you reach your destinations on time, then wind up in a comfortable bed.'

'That's a great idea, but why do you say imperfect?'

'Well, they seem to be good at certain things, but useless at others. Mine is great at food and drink, but if I don't plan the travel side meticulously, something always goes wrong.'

'So what is mine bad at?'

'How should I know? It's your guardian angel! You have to work it out for yourself.'

We found a great fish and chip shop where we took away a take away. Phil explained to me how good fish is for the brain, and that he ate it at every opportunity. I am not easily put off my food, however, so continued to shove it down my neck while getting the feel of the place. Newquay was very lively, packed with a crazy mix of typical family-on-holiday groups, surfing dropouts and young revellers. As usual, I was both delighted and horrified by the (lack of) clothes worn by modern women, which leave little to the imagination. Oddly, I don't remember such thoughts when I was their age: mini skirts were in vogue, yet my imagination worked overtime.

We only had time for one pint, in The Sailor's Arms, as sensible licensing laws meant all pubs had to close by half past ten except the 'over 25s' disco, which was actually surrounded by 15-year-olds and hardly our scene – at least, not that early in the walk.

The atmosphere on the way back to the B&B was not as relaxed as it had seemed before, with banter between the sexes harder and more serious, and there was a threatening feel to some of the groups of lads we passed. Passing a gift shop, I saw a T-shirt I just had to buy – it had a crazy drawing of a weird looking cartoon character on the front, with the word 'nutter' emblazoned beneath, seeming to sum up the situation perfectly. I was pleased to get back to the B&B, although there was still no sign of any other guests: were we the only ones?

It was gone midnight by the time my head hit the pillow, after a long and eventful day, yet the walk had not even started. I slept well until Phil's travel clock alarm went off at 6.00a.m., despite the fact that we had agreed to use my watch alarm, set for 8.00a.m. I tried to drift back to sleep, but found myself thinking about how it was I had set myself up to spend three weeks walking with Phil despite knowing what a challenge it would be – the wrongly set alarm was only a hint of what might happen.

YOU HAVE REACHED
GILMANS POINT 5685 M. A.S.L.

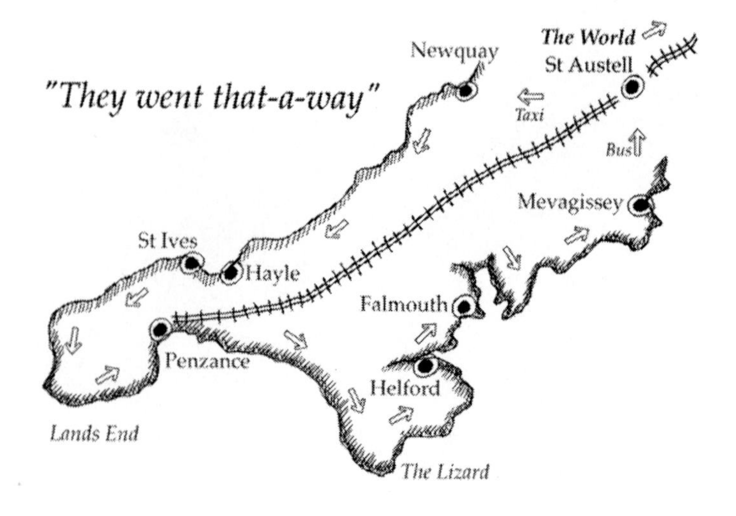
"They went that-a-way"
The World
Newquay
St Austell
Taxi
Bus
Mevagissey
St Ives
Hayle
Falmouth
Penzance
Helford
Lands End
The Lizard

Why? What? When?

I like to set myself, or just take part in, physical challenges. I will do pretty well anything which sounds interesting, having once taken part in a Horseless Horse Trial – the water jump being particularly challenging and superb fun. My friend and I, having climbed over the fence and jumped eight feet into thigh-deep water, pretended to fight each other. This was supposed to keep his young son amused, but I have often wondered since whether he was scared stiff that his dad was going to be beaten up and left to drown.

When I say I will do pretty well anything, this is completely different from saying I will do pretty well at anything. Most of my activities have been confused, shambolic muddles, and some have come very close to disaster, but I always have great fun doing them. My main reason for accepting such challenges, however, is that I enjoy seeing how far I can push myself, and get a kick out of burning off people who think they are going to leave me for dead. I hate people saying 'I'm too old to do that' and am tempted to reply with something along the lines of 'well, just get in your box now, and we'll screw the lid on later'. Assuming we only live once, saying you are too old to do something is effectively saying you will never do it again, which I find sad.

Given that I like to eat more than is good for me, particularly as I have a fairly sedentary job, I need plenty of physical action to keep in any sort of reasonable shape. Lynn prefers all-inclusive cruises in the Bahamas, or a fortnight at a Couples resort, to a week of hill walking, and she forces me to go along and indulge. At the end of such breaks, my stomach tends to resemble a beer barrel, and to get it back to anything even vaguely resembling the famed six pack requires a fitness drive, which only happens if there is a physical challenge involved.

A friend's philosophy is becoming increasingly appealing: 'Who wants a six pack when you can get more in a barrel?' He also claims that women like to look at six packs, but cuddle barrels, although I don't know how scientific his research was.

One of my work colleagues had a group of friends who liked to indulge their mutual enjoyment of hill walking, comfortable accommodation and good ale, particularly Theakston's 'Old Peculiar'. They had started modestly with a mixed programme ranging from one day walks in the South East of England to weekends in the Lake District, but graduated to such substantial footpaths as the Pennine Way and the Coast to Coast (crossing England from St Bee's Head in the Lake District to Robin Hood Bay in North Yorkshire). In keeping with their mutual interests, following a hard day's walk, evenings were spent in a pub replacing the liquid and energy used during the day, and nights in comfortable beds in hotels, pubs or bed and breakfast accommodation.

In the spring of 1995, my colleague asked if I fancied joining them in climbing Mount Kilimanjaro that autumn. They sounded like my kind of people, so I took up the offer, making the numbers up to six, including Phil. The trip was to consist of one week on the mountain, with a second relaxing at a hotel in Mombassa Beach. While my preferred approach is to do the bingeing first, then use the physical part to regain fitness, this was my first adventure with the group, and I was more than happy to do it their way.

During the spring and summer we went on several training weekend trips, which enabled me to get to know the other group members. Phil's oddball character, but basic generosity and willingness to do all the planning, room booking and other background tasks, soon came to the fore, and we became good friends. I soon discovered that one reason he had time to do so much for the group was that he had retired eight years earlier from the finance department of a London Borough, on grounds of ill health, despite being in his forties and one of the fittest people I knew.

He had been the union rep for years, helping members fight all sorts of injustice, but when it had been his turn, and the council had decided to make him redundant, no one had helped him. He fought his case single-handedly, including taking the council to court, where he acted as his own barrister against a vast array of top council officials and their legal team. Everyone involved, including some of those coming with us to Kilimanjaro, had told him he didn't stand a chance, should just go quietly, and offered no support.

Ultimately, he obtained a settlement that gave him a basic income for the rest of his working life, with a topped up index-linked pension to look forward to. He now worked a couple of days a week as a lecturer at various universities, spending the rest of the time indulging in his favourite pastimes of touring / sightseeing / walking / skiing and steam train riding. His situation was made even better by the fact that his brother worked for British Airways, which meant he could get extremely cheap flights worldwide. Wherever he went, he always sent a card to his finance department, expressing his appreciation of their good work, and reassuring them that he was managing to cope with his health problems.

Despite his clear final victory, Phil was still bitter over the way he was treated by the council, and particularly by his friends. I greatly admired his fighting spirit, which had enabled him to turn the threat of redundancy into the kind of deal most people dream of.

A particularly memorable trip was when we took on the Yorkshire Three Peaks Challenge, which requires climbing Whernside, Pen-y-Ghent and Ingleborough within twelve hours. At the time, one of my treasured possessions was a metal-framed rucksack, but when a magnificent thunderstorm approached on top of Ingleborough, the others discussed whether it was a good thing to have metal strapped to your back with lightning striking all around. I tried to ignore them by convincing myself that the manufacturers wouldn't make metal-framed rucksacks if they were good lightning conductors, and if they did, the Government would ban them, so they must be safe.

Nevertheless, I decided not to put it to the test if I could help it, so set off on the descent at a very impressive rate, ridiculously holding my rucksack at arms length as if that would stop me being involved if it was hit. A string of encouraging calls and comments from my 'friends' showed their appreciation of my plight: 'faster, it's catching up on you; show your metal; wow, John, you're like greased lightning; you always were a bit flash; is it something we said? Is that a lightning streak in your trousers?'

This was halted at one point when a bolt of lightning struck very close to me. My knee-jerk reaction was to throw the rucksack about twenty feet in the air, and for a few seconds they, by now some way behind me, thought I had been hit, but burst out laughing when they realised the truth.

In addition to the training weekends, we were encouraging each other to get in shape by jogging, going to the gym and playing tennis, so by the time the trip came round we were all pretty fit. We flew to Nairobi and made our way to Moshi, the nearest town to our chosen Marangu route up Kilimanjaro. We made our final preparations and worried about how it would work out.

At the park gate next morning, the tourism manager explained that it takes three days of continuous uphill walking to reach Kibo Hut, at 4703 metres, about 15,500 feet, with the final push to the 5895 metre / 19,340 foot Uhuru Peak summit taking place early the following morning. The six of us were accompanied by three guides, two cooks and fifteen porters, who carried all our gear apart from daypacks.

Day One's walk was scheduled to take three hours, and we kept religiously to the target pace set by the guides, who constantly reminded us to go 'polé polé' (slowly, slowly) so as to conserve energy, and because it is important to gain altitude slowly if you wish to minimise its effects. We made our way through tropical rain forest, on what started as a road, but slowly deteriorated into a cinder track, a four-wheeled vehicle route, a good walking path, and ultimately a rough walking path, which was as bad as it got the whole time we were on the mountain, and suited us perfectly. We would have been exceedingly upset if there had been a four-lane highway to the top, or one of those high quality stone and wood paths found on an increasing proportion of the popular walks in Britain.

The cooks and porters were far too quick for us, despite the latter each carrying four rucksacks tied in bundles of cloth and placed on the top of his head. They used one hand to keep the pack in position, and the other to push branches or undergrowth out of their way, help climb over tree roots, swat flies, and do whatever else they wanted to do with it. While we 'polé, poléd', the porters went up to the first night's hut, back down to the start and back up again with another load, and the cooks started to prepare the evening meal and ensure we had a cup of tea waiting for us when we arrived.

At first there were quite a few local people around, all very friendly, with the children very grateful if we gave them a few sweets, but soon we left them behind and focused on the walk. We were in great spirits, listening to, and trying to spot, the birds and monkeys high overhead in the canopy. An average of seventy walkers start each day, accompanied by a couple of hundred guides, cooks and porters, so there is a fairly constant stream of cooks and porters scooting past in both directions, and occasional liaisons with other groups.

When we started there was a group of twelve very fit looking mid-twenty year olds, all wearing T-shirts with marathon logos and other proclamations of superb fitness. They soon left us in their dust, and we did not see them again until we reached the night one huts at Mandara and discovered they had done the walk in one hour forty-five minutes. The opposite style was displayed by a coach load of Japanese, whose ages looked to range from early twenties to mid seventies, who took four hours, having stopped constantly to take pictures.

An aspect that had not occurred to me was constantly meeting the people who were now on their way down, but this proved to be fascinating. Well under half had made it to the top, and I didn't need to ask which ones they were: beaming smiles and inner satisfaction, or long faces and self-doubt told their own story. Most looked in fine physical shape, but there was the occasional twisted ankle or heavily bandaged knee.

Night one's accommodation consisted of a collection of huts made out of whole tree trunks. Twenty were triangular shaped, with an entrance and three bed spaces at each end, so could sleep up to six people. Meals were taken in a large wooden building with long benches, mainly occupied by Germans singing drinking songs, which was very strange because I couldn't remember seeing any on the walk. We soon joined in, and had a pleasant evening, apart from most of the others complaining that the food wasn't what they were used to. Before turning in, I spent half an hour doing one of my favourite activities: stargazing. They were fantastic since we were far away from city lights and other atmospheric pollution.

Day two started with our guides bringing us tea in bed, and after a leisurely breakfast, we set off. Five hours was the target time, and we continued to keep to the guides' pace, but with some dissension in the ranks – jealousy and rivalry had set in regarding the super fit one-hour forty-fivers. Why is it that, even though we had all day to get there, were all going to end up at the same place, and the guides had told us that 'polé polé' was the way to do it, some people still want to compete? I was soon voted down, however, and had to be content with beating the Japanese.

We soon climbed out of the rainforest and onto a massive plateau of open meadows and moorland on long, gently sloping hills. Things were going smoothly, we were well into the walk, felt fit and smug following our summer training, and were gently ticking over at an easy pace. The second night's camp at Horombo, 3720 metres / 12,300 feet, was reached comfortably, although the altitude was beginning to have some effect. We had taken the standard five hours, Superfits three and a half, and the Japanese seven.

Two of the group snored heavily, and Phil had been sharing a hut with them. Over breakfast on day three, he started to complain about the lack of sleep, so when he began to drag behind as soon as we started walking, I assumed it was because of that, and that he would pull through once he got going. Unfortunately, that was not to be.

Phil rapidly lost energy, started to look quite ill, and found it very hard to keep going. I tried to help by carrying his daypack and letting him lean on my shoulder, but it was obvious he was deteriorating fast. We made it over a large rock outcrop to discover a mile long gentle descent ahead of us, followed by an even longer gentle rise which we were informed led ultimately to the third camp. At first Phil improved on the descent, but by the lunch stop at the bottom of it, he was really suffering, with no energy and complaining of stomach cramps. Despite having organised the trip, it being one of his long term dreams, and all the training and effort he had put in, it took virtually no time for our head guide to persuade Phil to give up and turn back. As the remaining five pressed on, we could see Phil, with one of the guides, very slowly making his way back up the slope towards the outcrop.

Day three camp was at Kibo Hut, which we made by early afternoon, although progress was rapidly becoming more difficult as the altitude took increasing effect. We had to wait until 1.00a.m. the next morning before setting off on the final push for the summit. Virtually everyone had a headache, with the Superfits struggling more than most, two of their party having dropped out that morning. The Japanese arrived in time for tea, looking impressively good. We all tried to sleep, or just lay there, trying to forget how tough the final part of today's walk had been, and to keep positive about what lay ahead.

I heard three explanations as to why the final assault takes place in the middle of the night. Firstly, if people could see how steep and dangerous it was, they wouldn't go. Secondly, reaching the nearside, lower lip of the crater, Gilman's Point, at sunrise is incredibly spectacular. Finally, at night, the scree is frozen, so people don't keep sliding back every step they take – this is the version I believe.

At last, apart from another three of the Superfits who had dropped out due to severe headaches, we set out for the summit. Initially a long snake of some sixty people followed a well-trodden path zigzagging up the scree slope. Each little group had its guide, who, as ever, set the pace, which was now very slow, yet took all my energy. We just kept plodding away, in extreme cold and dark, for what seemed like eternity, very grateful for the ski gear and torches the porters had carried all the way from the park gate to Kibo Hut. I soon put my torch in my pocket, as I found it easy to follow other people's lights and preferred to have both hands free.

The remaining Superfits started to pull ahead, while other

groups dropped off the pace at the back, presumably led by the Japanese. Progress was becoming extremely difficult, but I was determined to hang on in there, so stayed tucked in behind our head guide following his every move. By now only three of our group were with us, but the other two's torch lights could be seen below, showing they were still making headway up the vast rock slope, and obviously not going to give up easily.

Watching other people move incredibly slowly brought home the pace at which we were now all moving. As we gained height, breathing became steadily more laboured and the effort involved in keeping going grew relentlessly. At one point, I made the mistake of sitting down for a rest: the effort of standing up again left me at least as tired as when I had sat down. Rests were to be taken standing up, ideally leaning against a rock slab, failing that, bent slightly forward with hands on thighs.

It was one of the strangest feelings I have ever had. I felt as fit as I had ever been, and was gulping in vast quantities of apparently wholesome air, yet I was moving at a pace unlikely to qualify for the final of the over-nineties Zimmer frame marathon for beginners, and it was utterly exhausting. Consciously, I knew it was the altitude having these effects, but my body and mind were finding it incredibly hard to accept.

The final couple of hundred feet to Gilman's Point consisted of scrambling up a solid rock face. This was a mixed blessing: the rock was firm (unlike the thawing scree below) but some of the ledges were extremely tiring to force yourself onto at that altitude. Finally, I hauled myself over the last ledge, virtually crawled up a gentle slope, and flopped down to look over the lip of the crater, which was quite shallow and contained several snow drifts, and across to the other side half a mile away. Dawn was not far off, and I sat and watched the sky slowly brighten, extremely pleased with myself. The plains of Africa gradually appeared, with mist drifting around, and the sky above slowly turning blue. Then the sun climbed over the horizon, the temperature rapidly rose, the mist disappeared, and a new day had begun.

Now came the bad news: the top was on the far side of the crater, only a few hundred feet higher than this side, but a two hour slog round the rim given our current condition and the high altitude. Three of us had made it that far, the others still struggling somewhere below, but with the scree having thawed... Our guide pressed for a decision: were we stopping here, or pushing on to the other side? If we were going on, it had to be soon.

We had been warned about the effects of altitude on the body, and had certainly experienced them, but no one had mentioned its effects on the mind. Logically, there was no way that we would turn back having got so far. We had trained for eight months, spent hundreds of pounds on the trip, and set ourselves the target of reaching the top. It didn't matter whether the other side was only a bit higher: it was higher. We hadn't come all this way and put in all this effort to get *nearly* to the top. If you are going to bother to climb a mountain, it is the top that counts.

One of the other two turned to me;

'What do you think? This will do, won't it?'

'Yeah, the other side doesn't look any higher to me.'

'What difference does a few feet make anyway?'

'It seems to go down at least as far on this side as it goes up on the far side.'

'It's probably a mistake, and this side is higher.'

'So we stop here, agreed?'

'Okay, agreed.'

While we had this conversation, the third member of our little group had been busy trying to fit a new battery into his camera. It was one of those B shaped ones, would only fit one way, and at normal altitude he would have seen immediately that he had it the wrong way round. Here, he was bashing the end of it with a small piece of rock trying to force it home. We turned to him for his decision. He didn't seem to even consider the idea of stopping: he had come to reach the top, and reach the top he would. Normally, I am sure this would have made us two rise to the challenge even if we had previously decided to turn back, but now we looked at him as if he were mad.

Consequently, our group's sole representative set off round the crater lip with the guide, while the two of us sat for a few more minutes watching the sun rise over the African plains, then started our descent.

As soon as we started down, everything became incredibly easy. We had gravity on our side, so the effort was much less and the lack of oxygen less of a problem. We were able to slide down the, now loose, scree and were soon scree running, taking massive strides at jogging pace, with half the mountain seeming to be sliding down around us. Probably very dangerous, and an environmental no-no, but a tremendous thrill.

Very shortly, I began to regret my decision to turn back. As we rapidly descended, the altitude effects disappeared and I realised what I had done. By the time I reached Kibo Hut, I was livid with myself. Ignoring everyone else, I threw my gear into my rucksack and headed off down the mountain at a ridiculous rate. By 11.00a.m. I was back at Horombo huts, our stopping place for that night. I started to blame the guide: he should have known how important it was to me and encouraged me to continue. It was still only 11.00a.m: even if I had taken another five hours to do the return crater trip, I could have been back here by 4.00p.m., hours before it got dark.

I found Phil, who was on the mend but still very weak, and we swapped tales and commiserated with each other. The Kili Killer had hit him with a vengeance half way back to the day two camp, and he had spent the rest of that day and night 'firing from both ends', getting through six pairs of shorts by morning despite spending the whole time within fifty yards of the toilet block. Only Phil would take that amount of gear on such a trek, and only Phil would use it all.

Nevertheless, he was very annoyed at his failure to reach the top, having put so much effort into planning and training, blaming himself for not being more careful with what he ate and drank. He accepted my altitude sickness excuse without criticism, which made me feel even worse, and when he agreed that the guide was more to blame than me, I suddenly realised that it was rubbish. I had to be responsible for my own actions – if I had been successful, I would have claimed the credit, not said it was due to the guide.

Soon other walkers began to arrive and we heard their stories. Neither of the members of our group who had dropped off the pace made it to Gilman's Point, but they were both pleased with themselves for having given it their best shot, which made me feel even worse. None of the Superfits had made it that far either, but all bar two of the Japanese had made it all the way, several of them not reaching the top until well after midday. There were some very clear, strong messages here about teamwork, stamina, determination and pacing yourself. No one pointed them out – they didn't have to.

During the evening and night I did a lot of thinking, and by the morning my mind was made up: Instead of going with the others to Mombassa Beach, I would be staying here for the second week to do the climb again, but this time I would make it.

When I told the others of my decision during breakfast, I received a mixed reaction. They were unanimous in their complete lack of faith in my ability to do it, but some were highly impressed that I was prepared to give it a try, while others thought it was clear madness. Phil was now extremely weak, still unable to eat, and struggled to give the idea of anyone even contemplating going straight back up any credence whatsoever. He periodically burst into fits of laughter, or gave me a quizzical look while repeating that I just couldn't be serious. I knew he was rather prejudiced by his recent propensity to puke, but even so felt he could have shown some respect.

These reactions strengthened my resolve, and on the final day's descent I became ever more determined to carry it through. When we reached our hotel in Moshi, I told the manager of my intention and requested her advice and assistance, which she gave very willingly, immediately determining to help me succeed. There was an excellent guide she knew well who had a reputation for getting people to the top, and she would see if he was available: he was, if we started the day after next, so I booked him. Phil finally accepted that I really was going to do it, and went into mild shock, muttering to himself how I was even more of a nutter than he had thought.

I asked the manager if anyone else had attempted the same plan, and she said that she had worked there for twenty years and had never heard anything like it. Word soon passed round, and during that evening people kept coming into the bar of the hotel, pretending to look at the paintings on the wall, but nudging each other and pointing at me. I'm sure several bets took place, and that the smart money was not backing my success.

The others flew to Mombassa Beach the next day, in accordance with our original schedule, and having agreed to phone Lynn to explain the situation to her: even I was not mad enough to do that myself. Her reaction was: 'Trust him' which proves she understands me better than anyone else, probably including me.

I had one day to prepare myself for what just had to be an extremely tough week. I felt good, but had no way of knowing how much the first week had taken out of me, nor whether I would be able to cope with an early return to altitude. Other than wash my clothes and have a couple of showers, there didn't really seem to be anything I could do. After a few of hours lounging around, I went for a walk.

Kilimanjaro is often shrouded in cloud, but there it stood, proud against the sky and towering contemptuously above me. I felt awestruck and began to have self-doubt. I had already tried and failed, was it sensible and right to go back so soon? I don't believe in gods of any description, but still prayed to the mountain god to accept my attempt as a personal test and not a challenge to the mountain, and to lend me some of his strength to succeed. The feeling of malevolence passed and was replaced with one of peace, strength and encouragement.

I regained my faith and determination immediately. My lack of belief was severely shaken: who was I to decide whether or not Gods exist, and if they do, how many there were, and in what forms or guises? Obviously, psychologists could easily explain my initial doubts and subsequent rediscovery of immense confidence, but I had felt a very strong, external force change its attitude dramatically in my favour, and I certainly wasn't going to look for rationalisation.

During the first day of my second attempt, I almost overtook a girl who was backpacking round Africa and had decided the previous day that it would be fun to climb Kilimanjaro. Her pace made her look as though she had invented 'polé polé', and she would certainly have given the Japanese a dawdle for their money.

Having spent half an hour or so on idle chat, I had just decided to push on ahead when I realised she was, possibly literally, a God send.

If I had learnt anything from the previous week, it must surely have been that the guides were right to emphasise the importance of 'polé polé'. Here I was with Miss Polé 1995, and waving her goodbye! If I was really determined to succeed, the obvious thing to do was to stick right there with her, let her set the pace, take it easy, and save everything possible for the final push. We spent over five hours getting to night camp one, Miss Polé receiving many approving glances from the guides and other locals.

On day two, Ann developed blisters, and really began to struggle, but I was happy to keep with her, and we still kept to just outside Japanese pace. On the morning of day three, she decided to stay at that level for another day and night, to give her blisters a chance to heal and her body a chance to acclimatise, which is what the guides recommend. While I knew I would miss the company, and would now have to set my own pace, I was pleased in that I would now be able to focus completely on my own success. It really did seem as though she had been sent to ensure I kept to the optimum pace for the start of my second attempt, and having done that job, was now being taken away so as to avoid any interference in the final stage.

All the time since setting out on this second attempt, I had been waiting for the previous week to catch up on me, but I still felt great. I spent a lot of time talking to my guide about the training they go through, whether the number of walkers was having any effect on the vegetation, and whether he had noticed any signs of global warming. He told me that the amount of ice and snow at the top had roughly halved in the last twenty years, and it was getting less every year.

Kibo Hut was achieved relatively easily, and I managed to get some sleep, despite having a dreadful feeling that last week must hit me soon. The night walk started well, but soon the altitude began to hit again. I think it was at virtually the same height on both occasions, say 17,000 feet, that the effects really became serious. Again, every step was a real effort and I had to force myself up the final climb to Gilman's Point. This time, however,

I was ready for the mental side, and already determined to push on no matter what false feelings of success assailed me. Probably because I was ready for them, no such feelings came, and my guide and I soon started round the crater towards the real top.

After half an hour of gently descending flat walking, we came to the final long ascent. There was only one guide and walker pair in front of us as we inched ever closer to the top, taking frequent standing rests and forcing ourselves to gulp for air. I still found it an incredibly strange sensation, knowing I was breathing far more quickly and deeply than normal, there was no obvious difference in the air I was breathing, and yet I was just not getting enough oxygen to supply my muscles so as to slowly crawl up a slope I would be able to sprint at sea level.

From Gilman's Point there had not been much snow to see other than that in the crater, but on this final stretch the snowdrifts were suddenly all around us, ranging up to thirty feet deep. We were still walking on solid rock, however, the final ascent being mercifully flat and smooth. Eventually, we made it to the very top, and were able to see Gilman's Point on the other side of the crater, clearly below us.

My feeling of success was far greater than I ever would have felt if I had done it the first time. Not only had I achieved the climb, but after failing the first time I had had the courage and faith to turn straight round and go back up. I felt an inner satisfaction that I had never felt before, and vastly increased belief in my ability to take on anything. There is no doubt in my mind that this was my finest hour.

Kilimanjaro had made Phil and me strong friends, with a great deal of mutual respect for each other. I felt sad that he had done nearly all the planning and organising, been going very well, but then had been stricken by the dread of all travellers. It took him a long time to really come to grips with my immediate return, having got so close the first time.

Many people just don't see why I went back, particularly when it would have been very easy to get away with saying I had climbed it the first time. The park authorities give a certificate to anyone reaching the top, but they have two versions. Both say

'...climbed Mount Kilimanjaro...' but one just has the words 'Gilman's Point' added. Only people who had actually made it to the top would know the difference, and only then if they had friends who stopped at Gilman's Point.

I had set myself a challenge, and I knew I had failed it, and done so without giving it my best shot. It meant little to me whether other people knew or not, and (the part which very few people seem to realise, even if they do see the previous point) if I had given my best shot, but still failed, I would not have been too badly upset.

I do like to set myself tough challenges, then have a really good go at achieving them, but if I allowed myself to get too bothered whenever I failed, I would either end up a very unhappy person, or restrict myself to things I knew I could achieve, which would effectively remove the challenge element.

Having enjoyed Mount Kilimanjaro, I joined the group on other great walking trips including to the Faroe Islands and Madeira. When it came to planning for 1999, however, the comfortable accommodation and good refreshment side seemed to dominate the walking, and the main event was agreed to be a week-long train-based tour of Eastern Europe's capital cities. Pathetic: it was supposed to be a walking club, not SAGA rejects. I decided to opt out and look for an alternative genuine walking break.

While Phil would be keen on a train-based tour of the local rubbish dump, and immediately signed up, he also recognised the lack of physical effort involved and was keen to do more. He suggested an attempt on the South West Coastal Path, which starts in Minehead, on the north coast of Somerset, and follows the north coasts of Devon and Cornwall west / south-west to Land's End, then east along the south coast to South Haven Point, near Bournemouth, Dorset. Phil's dad was born in Torquay, and he had visited many of the interesting places on the path in his youth, but not since taking up walking.

Phil pointed out that, at 965 kilometres (around 600 miles) long, the South West Coastal Path would make a good dry run for any future attempt on the Appalachian Trail, one of my pipe dreams, as I am sure it has been for thousands of people since reading Bill Bryson's *A Walk in the Woods*. Coincidentally, there would be a total eclipse of the sun in Cornwall on the 11th of August 1999, so we could include that as a bonus. I was convinced, but none of the others had both the time and inclination to join us, so we decided to go as the gruesome twosome.

Six hundred miles at, say twelve miles a day, is fifty days, or just over seven weeks. Both Phil and I are in the fortunate position of having that kind of time available for relaxation, me due to being a university lecturer, Phil due to his brilliant redundancy deal. Phil is not quite as fortunate as me because I have an extremely beautiful and caring wife, who soon informed me that she cared far too much for me to let me clear off with Phil for seven weeks, leaving her to sit indoors feeding the goldfish.

There were two obvious flaws in her argument. Firstly, we have no goldfish. Secondly, there is no way that she would just sit indoors. Lynn is a travel fanatic, who job shares at Thomas Cook so she can get access to cheap holidays, giving them back her wages to pay for her trips. I knew that she would be off on some cruise to Alaska, a Las Vegas sampler, skiing in New Zealand or 'flying down to Rio' as soon as she saw my rucksack back.

I thought I could persuade her to let me go, but knew it would require a deal of grovelling from me, including agreeing to several future trips for her, with me left at home milking the camel. I was also not sure I could stand being in the close company of Phil for seven solid weeks. Besides, I do like to spend a fair chunk of my spare time with Lynn: I might even miss her.

While I was puzzling this out, some friends I had met while skiing during an extremely tough conference in Jackson Hole, Wyoming, USA contacted me with a very strong request that we should visit them at their chicken farm in North Carolina during the summer. This opened up the possibility of spending three weeks in America with Lynn, developing a nice beer barrel, then a similar time doing part of the South West Coastal Path, converting it into a six pack. The Appalachian Trail passes through North Carolina, so I could even include a quick recce during the America bit.

Talking to Phil, he confessed that he was hoping to spend some of the summer in Greece courting one of his students. I told him to keep his kinky sexual habits to himself, but it all seemed to be fitting together: we would do our own things in July, then spend most of August trail tromping. I had to be back by Saturday the 21st of August in order to attend the Oval Test between England and New Zealand: I never have a moment's rest.

Lynn saw that if she did not act quickly her holiday in America might start to evaporate, so pre-empted the position by booking flights. I find her independence one of her most admirable characteristics, which is just as well, because there is nothing I could do about it if I didn't. We would return on Wednesday the 28th of July, so I could not really leave for the walk until at least the Friday. Our three weeks were beginning to be under threat, so we quickly agreed to start on either the Friday or Saturday.

Phil and I agreed to have a planning meeting in June, and he arrived with the relevant six Ordnance Survey Explorer Series maps and a copy of the book *The South West Coast Path 1999 Guide* by the South West Way Association. He had also decided that the best place to start would be Newquay, and that we should be able to make it to Land's End and round to Torquay, a total of three hundred miles, in three weeks, i.e. one hundred miles a week, say fifteen miles a day.

I am usually a very independent person, but having never been to Cornwall before, was happy to go along with Phil's suggestions. We both knew that our plans had to be kept highly flexible, since progress would depend on many factors including weather, fitness, injuries, available camp sites, and any interruptions such as the eclipse and general sightseeing – having gone all that way, it would be a shame not to spend some time looking at the scenery. Phil had already suggested the possibility of spending a day or two on the Scilly Isles – 'a top place: you'd love it' – and visiting Burgh Island either on foot at low tide, or by sea tractor at high tide – 'great old pub, called The Pilchard, with an art deco hotel used by Agatha Christie as the base for several of her murder mysteries'.

Being experienced walkers, we both knew exactly what to

bring once we knew what type of walk was to be undertaken. Reading the guidebook and examining the maps, coupled with Phil's local knowledge, made it obvious that we would be very fortunate to obtain accommodation every night, particularly given the rumoured two to three million people invasion during the eclipse period. Such an approach would require pre-booking, which would remove the flexibility we were keen to maintain. We were, however, confident that we would be able to obtain meals relatively easily, so agreed to take a tent, sleeping bags, knife / fork / spoon / mug / plate, but no cooking facilities. Neither of us owned a suitable tent, but my daughter did, and had said that we could borrow it, provided we looked after it.

We decided to erect the tent in my garden, to make sure we could, and check for holes. It was a dome tent, with two sets of thin poles which link together and are then pushed through tunnels in the outer tent fabric such that an igloo shape is formed when the ends of the poles are inserted in eyes in the fabric, and tent pegs strategically placed. The inner tent is then hooked on to rings in the outer tent, and a third set of poles inserted to form an entrance 'porch'. Experienced campers can do all this in three minutes flat, and we soon had it down to twelve.

The only problem with the tent was its weight, seven pounds, but Phil thought that was okay since I would be carrying it. Phil suffers from glaucoma, and takes eye drops to control it. These have several interesting side effects, one being slowing his heart beat, which means that, in a dreadful parody of the song by Billy Ocean 'when the going gets tough, Phil just stops going', so I had agreed to carry the tent to help lighten his load.

Another side effect of the eye drops is that Phil is always incredibly cold, tending to wear two T-shirts, a jumper and coat when everyone else is commenting on how hot the outdoor pool is. He has been known to wear eight layers of clothes, including jumpers and anoraks, when exposed on a mountain ridge. Which reminds me of an occasion when one of the group decided to risk exposure on a mountain ridge in the Lake District while attempting a peeing long distance world record over the edge, but failed miserably due to not checking which way the wind was blowing. He was left with egg on his face, or something.

Presumably for similar reasons, my dad always says, if you feel ill on a boat, before being sick over the side, spit.

Side effect number three is that Phil has to pass water frequently, which, on walks, can be problematic or entertaining, depending on one's viewpoint.

Having settled our plans, all we really had to do was protect the first three weeks of August from encroachment.

One of my nieces, Sally, was having a twenty-first birthday party on Saturday the 7th of August, and Lynn said she thought I should attend, in that way that tells me to agree. I was able to book a train home from Penzance on the Saturday morning, but due to the eclipse on the following Wednesday, was unable to book a return until the Monday morning. Phil was happy to go along with this, as he could use the spare time to nip across to the Scilly Isles.

In addition to his fanaticism about steam trains, Phil is somewhat keen on football, having watched a game at all ninety-two of the Premier and Nationwide league grounds, a feat he regards as the major achievement of his life. He supports Wolverhampton Wanderers (Wolves) one of the top teams in the '60s and '70s, with such household names playing for them as Derek Dougan and er... Since then, they have had a run of bad luck covering twenty-five years, but Phil is a loyal fan and has followed them all over the British Isles.

Phil lives in Wycombe and had never seen Wolves play there, but, in one of those quirks of fate which seem to happen to everyone else once in their lives, but to Phil every time he changes his socks, Wolves were to play Wycombe, on Tuesday evening, the 10th of August. 'I'll have to go. I can't believe it. Wolves at Wycombe! I'll miss the eclipse unless I drive back through the night, but this is much more important.'

In addition to knocking out Tuesday, it was fairly obvious that we would not be doing much walking on Wednesday, eclipse day, with Phil either on a train or sleeping after driving from Wycombe to Cornwall overnight, probably with a million companions for the greater part of the trip. Bang go another two days.

The final inroad into our time was when Phil phoned to say he had a great idea for how we could start – by joining a steam train trip from Paddington to Kingswear on *The King* on the Sunday. He sold it to me as giving me another couple of days to prepare after my America trip, but I also knew that, with a steam train involved, there was little point arguing.

Our three plus weeks were now down to five days walking, a five-day break while we did our own things, then nine more days walking. We decided to revise our target to Newquay – Penzance during the first five days, then set up camp in the Penzance area until we started walking again, whereupon we could set ourselves a target for the final week.

As soon as she heard of the delayed start, Lynn managed to reschedule our return from America to the Saturday, so I ended up slinging my walking gear together in a jet-lagged stupor:

Rucksack:
Large, black and multipurpose. Designed for use on plane trips, as well as more traditional backpacking, the straps could be folded and rolled inside a zipped compartment so they didn't catch on conveyor belts, bus doors etc. and it had grab handles on the side and top. My first act was to tactfully dump the America stuff on the floor, causing a rather strained atmosphere until Lynn decided to go shopping, slamming the front door on her way out.

When used as a standard rucksack, it is beautifully comfortable, with vast amounts of padding round the straps and all the nice touches like hip and chest belts. Given all these great features, the drawback is obvious: it weighs more than it should. Phil does not think this matters for the, again obvious, reason: he isn't carrying it.

Rucksack liners:
Specially designed waterproof liners are available which are contoured to fit the rucksack so as to keep all contents clean and dry. They cost at least five pounds, so I had dustbin liners instead.

Hat:
Important for keeping one's head cool on hot days, and warm on cold ones. I use a baseball style, the peak helping keep the sun out

of my eyes and making it harder for autograph hunters and paparazzi to recognise me (when I'm rich and famous due to the millions of people who buy this book, tell all their friends about it and cause a film and mini comedy series to be made by popular demand).

Trousers – 3 Pairs:
Pair one were standard walking trousers, but with nine pockets, seven with zips, including one near the outside of the left knee. Excellent for handling all the stupid odds and sods one tends to want fast access to when walking: cash, cards, keys, energy bars (I love toffee or scroggin, which is Australian for a mix of fruit and nuts) hanky, bits of string, used bus / train tickets, pen and paper, maps, route books, interesting fact sheets on anywhere you have visited or might visit that day. The only problems are that it takes five minutes finding which pocket the particular little odd sod you want is hiding in, and that having so much space encourages you to keep all sorts of rubbish in your pockets instead of throwing it away or packing it properly. Consequently, I end up looking as though I have incredibly well developed, but odd shaped, thigh muscles, and a nasty growth on my bum.

The second pair were walking shorts, very similar to the others, but without legs, for those long hot summer days you read about. Incredibly thin, light, but strong.

Pair three, a combo pair bought by Lynn on a recent trip to Beijing. Long legs for when it's wet or walking through rough vegetation, but with zips near the top of each leg to allow the wearer to convert them into shorts when the sun comes out. The leg bottoms are expandable to facilitate removal without taking boots off. These are worth their weight in camel dung just for the brilliant reaction from people sitting opposite on the train when you 'do it'. When in a particularly silly mood, I have been known to walk with one leg on, the other off.

Shirts – 3:
T-shirts. One a frighteningly bright orange / yellow, included for its signalling qualities in any search and rescue exercise we might happen to be involved in. The second black and made out of some incredibly sexy see-through mesh fabric which would have

driven the girls wild if worn by Mel Gibson. Number three was white, but with great designs on front and back depicting the Jamaican bobsleigh team, made famous in the film *Cool Running*.

Underwear:
Under where? You don't want to know. If you do, you sound like my kind of girl, so send your name and address, with pictures, to Scanty_Panty@bodystudies.ac.uk. If you do, but are male, get a grip on *yourself*.

Socks:
As in all good army films, where our hero is kitted out: socks, pairs x 5, size 9, walking, for the use of. What more can you say?

Lots, actually. If you want to be a happy walker, you must 'treat your feet'. Having the best possible walking socks, in a variety of thicknesses and styles to enable variation depending on walking surface and foot size (feet can expand when hot).

I had two thin pairs, two thick pairs and a pair called Le Double. These were 'fabrique en France', and consisted of two layers of material, which are designed to slide over each other, one moving with your boot, the other staying with your foot, which supposedly prevents blisters. I cannot vouch for their effectiveness, because I never get blisters – at the time I thought Phil didn't either.

Footwear:
'Treat your feet'.

Even now, I feel guilty writing that due to the new depths of hypocrisy it plumbs. I do treat them – like pieces of wood that belong to someone else, or slaves that dare not complain in case they get even worse treatment in future. They probably wouldn't be surprised if I cut one of their toes off just to show them who's boss.

I had two pairs of things for bunging on my feet. Primarily, I had my wonderful, beautiful, God I love you so much, 'Approach' boots.

A couple of years earlier, I had almost been killed when, on the Three Peaks challenge (Ben Nevis, Scaa Fell Pike and Snowdon, all in twenty-four hours), I started to slide down a grassy slope towards a sheer drop, due to lack of grip on my boots. I blamed the boots for letting me down at such a vital time, when the truth is that they had been losing their grip for ages and I had done nothing about it.

Shortly after that, on my way to do the Coast-to-Coast walk, I suddenly realised I was starting a 200 mile walk with boots that, although very comfortable, basically had no sole. I had arrived at Euston station with twenty minutes to kill, so went in search of a sports shop, which I found, and enquired as to modern walking boots.

Being a superb salesman, the guy completely sold me on the line that these 'Approach' boots were a revolutionary new concept in outdoor footwear. They were a hybrid, combining the strength and support of traditional walking boots with the lightness and comfort of trainers. Called 'Approach' boots because they were designed for people tackling Everest to wear on the approach walk to the climbing part.

I knew this was all sales flannel, but it had the desired effect of getting my imagination running, and I bought it. I guess the fact that I now only had six minutes before my train left, and a desperate need for grip, had some influence too.

They were unbelievable. I wore them for the whole ten days of the Coast-to-Coast walk without a single twinge or blister, and had hardly taken them off since. I had walked, run, danced and even lectured in them, and they had always been brilliant. They even looked smart, being a fairly dark brown colour and made of quality looking material. Do you get the idea that I liked them?

Knowing I could do anything and go anywhere in my marvellous Approach boots I didn't spend much time on what other footwear I took. I had a pair of walking sandals which I had found comfortable and cool on short walks, so decided I would take them, mainly for wearing in the evenings around the camp site or in pubs.

Waterproofs:
Another vital piece of gear. Getting cold and wet on a walk is no fun, and can be dangerous. My trousers were eight sizes too big for me: I am yet to meet, and definitely don't want to, the person

they were designed for. To prevent myself tripping on them, and stop them dragging in the muck and bullets, I tried rolling up the trouser legs but it did not work since they rolled straight back down again. What worked was pulling the waist up to my armpits, then either leaving it there if I was cold, or rolling down the spare fabric until I reached a more traditional level. Elegance aside, they were highly effective, and included zipped expansions at the bottom of the legs for easy donning and doffing without having to touch wet muddy boots.

For several years, my only waterproof top was a very thin, bright blue and mauve one designed for protection from summer showers. During a walk in Scotland earlier that year, I almost died from exposure having pushed on to the top of a mountain when a snow storm hit and all but one of the others, despite having quality four season jackets on, had turned back. This convinced me of the need to buy a proper mountaineering quality waterproof jacket, and uncharacteristically, I had done so.

I didn't fancy lugging my new jacket around for three weeks, so decided to take my old thin one.

Eye gear:

I am very shortsighted, wear contact lenses virtually the whole time I am not asleep and glasses for the rest. When on a walking trip, I make sure I have my glasses, in their case, sun glasses in their case, and at least two pairs of contact lenses, with their cases, cleaning solution, which I take in a small bottle so as to minimise weight, and a small makeup mirror. The mirror was bought for me by my daughter Claire, who is also rather bat-like in the eye department, so understands my needs. Nevertheless, I am always concerned that someone might see it and leap to the wrong conclusion: I deny wearing makeup, and there is no photographic evidence that I am aware of.

Toiletries:

Standard set, but travel versions: toothbrush, paste, soap, shampoo, the latter three being the small size ones found in hotel rooms, saving space, weight and cost, and a ludicrously small, six by ten inches, travel towel which acts like a sponge.

First aid kit:
Special lightweight walkers kit, received as a highly thoughtful Christmas present a couple of years previously. I had no idea what was in it, nor how to use whatever was.

Being a six foot fourteen plus stone ugly brute, I am not allowed to confess to being worried about being attacked, so the reason I took a personal attack alarm must have been to summon assistance if we managed to get ourselves injured or in danger, or completely lost. This helped me cope with the visions I had of slipping over the edge of a cliff, coming to rest on some hidden ledge, then taking two weeks to die of thirst, exposure and 'Phil', due to being unable to attract the attention of potential rescuers.

Odds and sods:
One hundred pounds cash, credit cards, driving licence, writing paper and pen; watch with alarm; house key, for the not unlikely situation of Lynn being off on a cruise when I got back; handy bits of string; sewing kit.

Maps etc.:
Phil would be bringing the maps, but had left *The South West Coast Path 1999 Guide* with me so I could pre-book some accommodation, which I had very nearly remembered to do. I had my waterproof map holder and a compass.

Ski pole:
I used to think only wimps used sticks when walking, but learnt on the Kilimanjaro trip, where they were compulsory, just how useful they are. They take thirty percent of the force away from the leg joints, can help with balance, and are useful for fending off wild animals, savage dogs and Phil.

My walking stick is not actually a ski pole, but that is what it looks like. It telescopes so as to fit inside my rucksack, being locked into the extended position when in use by twisting the inner and outer sections of pole in opposite directions. It has a wrist strap to avoid loss if dropped, and a small basket at the bottom, similar to those on ski poles, to prevent it sinking too far into the snow / mud / quicksand or whatever friendly surface I happen to be walking on.

Utensils:
Although we had agreed to each take a knife, fork, spoon, mug and plate, I only actually took my camping penknife, a dessert spoon and a mug, reckoning that they could be adapted for all purposes. I also had my Sigg water bottle, which is incredibly strong and light.

Night things:
No, I didn't take a nightie. I did take a tent, sleeping bag and ground mat. I meant to take a torch, but couldn't find it.

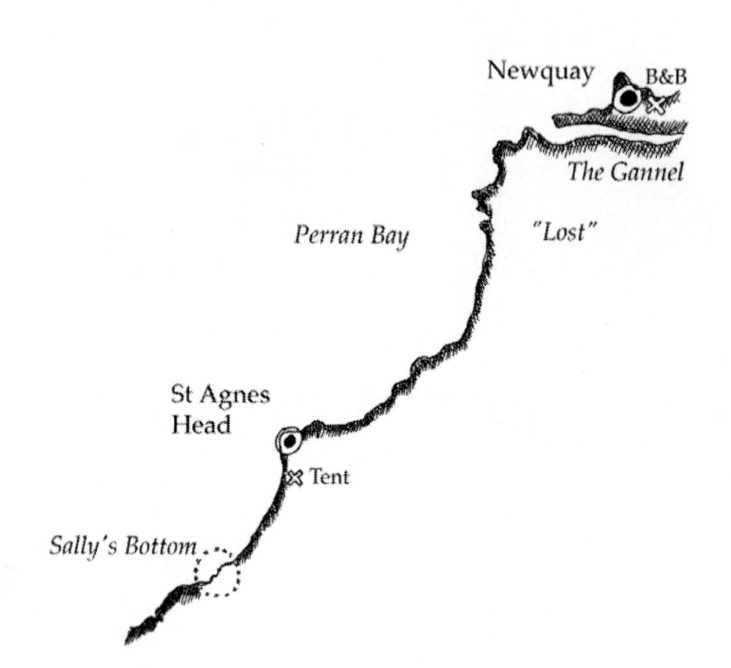
Newquay
B&B
The Gannel
Perran Bay
"Lost"
St Agnes
Head
Tent
Sally's Bottom

Bad Moves and Good Fortune

When we entered the breakfast room, we were surprised to find ten people already tucking in: where had they been last night? Instead of the typical self-service cereals bar, we were waited on by the owner, which was a nice touch, but rather disappointing when he only gave me two spoonfuls of bran flakes in the bottom of a large bowl. I toyed with using the 'excuse me, this bowl is dirty' line, but thought better of it. In any case, one of my objectives was to lose weight. On Saturday night I had weighed in at fourteen stones six pounds and my dream target was thirteen stones. Nevertheless, I was very relieved when I noticed the piles of eggs, bacon, mushrooms, tomatoes, toast... which formed the main course. One likes a good breakfast when starting a long day's walk.

At 9.50a.m. we, at last, set off, retracing our footsteps of the previous night. Our rucksacks were heavy, but not too bad, and we were soon back in the general holiday bustle of Newquay in early August. The family groups were out in force, but there was a marked lack of surfing dropouts and young revellers, presumably still recovering from the previous night's activities, when they could be setting out on three week's glorious walking. We received a mixed reception from those we passed, including a chorus from one group of 'I love to go a-wandering' which left me singing it for the next hour.

Having given Phil sixty pounds for my wonderful steam train ticket, I was already short of cash, so attempted to withdraw some from a hole in the wall.

PIN numbers always defeat me – I just can't remember four digits, which is a bad confession for a finance lecturer to have to make. At one time, I had a PIN, which I tried to remember as 'late back from lunch', the only problem being I couldn't remember what time I had gone for lunch, nor how late back I was! Once, a beastly machine decided it was time for its lunch, and swallowed my card when I guessed wrong three times in a row. A recent birthday card depicted a hospital reception with two queues: one for total amnesia, the other for 'forgotten PIN number only': I can't remember which queue I was in.

You will have guessed that I had no joy on this occasion either. My ghast was somewhat flabbered, since I had put special effort into remembering my PIN, given its likely importance on the walk. I was sure I had the number right, but after two attempts there was no way I would risk a third. It subsequently transpired that I was using my own card but Lynn's PIN. For some reason lost in the mists of time, I only know her PIN, and normally use her card. We had just been sent new cards and I had picked up mine by mistake.

Phil had brought an impressive wad with him, and had my sixty pounds on top, but still did not want to miss the opportunity of rubbing in his ability to remember four digits, so took out even more. What a ~~mug~~ good man!

We continued through Newquay and on to Fistral beach, which is one of the top surfing beaches in the UK, and where a work colleague had been almost drowned when caught in the undertow and unable to fight his way out – luckily a large wave took pity on him and dumped him onto the beach. An international junior championship was being held, but the waves were 'all fiddling and small' and the real action seemed to be stallholders attempting to sell poor quality T-shirts for fifteen pounds. There were too many slim, fit young men around for my liking, so we pressed on.

The South West Coast Path 1999 Guide made it very clear that the best way to cross the Gannel River to Crantock Beach was by the Fern Pit Ferry. This early on our walk, we had not bothered or got our act together sufficiently to read it yet, so headed instead for a tidal bridge i.e. one only usable at low tide. As we approached, a local informed us it would be two hours before the bridge was clear (the guide book had the tide table in it too) and suggested a two mile detour further up river, so off we went, very casually adding four miles to the day's mileage.

The Gannel winds through a several-hundred-yard wide

sandy valley, flanked by a mixture of woods, fields and beautiful houses with well-kept gardens. As the tide retreated the sand flats were uncovered, providing brunch for hundreds of wading birds. A large number of walkers, many with dogs, were enjoying the beautiful sunny morning, and on the opposite bank there was a group of horse riders, splashing along the shallows. We felt great, looking forward to three weeks of such idyllic environments, and relaxed into a steady pace, amusing ourselves with gentle banter.

Ever since leaving the B&B, Phil's tendency to pack bags incredibly poorly and leave zips and straps undone had been on display. Bits and pieces were constantly dropping all over the place, and he had to keep stopping to pick them up, often dropping other things in the process. During the train journey, in addition to the walker's compass and half-eaten toffee bar at Reading, he had dropped his credit cards at Didcot, where, luckily, one of the steamies had spotted and returned them. So far that morning he had dropped and retrieved his water bottle four times, his ground-mat, various sweets, wallet, and several articles of clothing. I was used to this habit, so initially took it very light heartedly, but after a while it began to irritate and I started to get sarcastic, then annoyed, and finally a bit ratty.

Suddenly I realised I was no longer carrying the map.

'I must have left it where I took my shirt off to get my tan going about a mile back.'

I decided to walk back to look for it, partly to get away from Phil's glee.

'Oh dear, oh dear. I drop things all the time, but at least I pick them up. You said you would do the map reading, now how are you going to do it? Hope your photographic memory's in working order. Do you know it like the back of your hand, or are you going to try palm reading?'

I told him to go forth and multiply, and set off. Without the pack on, it was an even more pleasant walk, and I couldn't help being in a good mood, even though I was having to redo one of the extra two miles each way, making a total of six extra miles for the day. The irony of the situation wasn't wasted on me, and I had a few quiet chuckles to myself, which worried some passers by, one going as far as to pick up her Rottweiler to protect it.

I had no luck, so returned empty handed to find Phil lying on the sand half asleep. We decided to continue by guessing the route, trying to remember as much of the map as possible. Every cloud has a silver lining: we were forced to open the wonderful guide book, to discover its rich vein of information, and how useful it could already have been. We vowed to make good use of the guide from then on.

Eventually we stumbled across Crantock, which is on the route-ish. We spotted the general store, so while I ordered some cheese and ham sandwiches, Phil asked the pertinent question;

'Maps? Yes, we keep a complete range so as to provide a quality service to people like you.'

'Great, we need *Explorer 104*.'

'Err, now let me see, should be here somewhere. Oh dear, now I remember: that's the local one – we always run out of that over the weekend. We've got several copies of 105 and 106, though.'

'OK, I'll take a 105 and cut the top off.'

The assistant got as far as to place a copy in a brown paper bag before registering that Phil had to be joking, looking up to see his deadpan expression, and sheepishly putting the map back in the pile. Later, Phil confirmed that he would have bought it and borrowed a pair of scissors to slice a strip off, just so the guy could spend a couple of weeks telling his mates about it – until one of them pointed out that the joke was on him.

While I distracted the staff, not that they seemed to care, Phil sneaked a look at another map, which covered a short section of the route, and armed with that knowledge we managed to find our way through some sand dunes and a golf course to Holywell. It had become a very hot day, and we were both low on water. I asked a friendly shopkeeper if she would very kindly fill my water bottle, and she agreed. I said:

'Oh, I've actually got two.' She agreed to fill both.

'Oh, and so has my mate.' She gave a sympathetic, almost conspiratorial smile, and agreed to fill them all.

Even I felt guilty, so I bought an ice cream while Phil had a bit of a chat with her about our intentions and progress so far. It was just gone 3.00p.m. so in five and a half hours we had advanced a measly seven miles. Due to our starting two miles the other side of Newquay, detour round the Gunnel, my searching for the map, and general wandering over the dunes, Phil had actually done thirteen miles, and I had done fifteen, only three of them on the official route. Plus I had lost the map. In all, he said, it was probably about par for our walks. For some reason, Mrs Holwell-Shopp seemed to doubt our success, but wished us well and pointed out another shop, which sold maps, including *Explorer 104*.

Overlooking Perran Bay, Phil, being a keen photographer, stopped to take a picture. He had already used three rolls of film, preferring to snap away constantly, in the hope that some good photos will result, which they often do. A very good way of ruining Phil's day, even if the sun is shining, birds singing, and there is peace and goodwill on earth, is to deprive him of this pleasure. He went to take a picture, only to discover that his camera's batteries were missing.

'I can't understand it. They were there before, but now they've gone. They must have fallen out somehow, the cover must have come loose.'

Just before having a go at him, I remembered the map I had forgotten. Since you can't see a photograph of it, I will describe the scene for you.

Two miles of flat sand lay stretched in front of us, with rollers breaking and washing onto the beach, which, with the tide now fairly well out, was several hundred yards wide. About a hundred people were in the water, paddling, swimming, playing ball and diving through the waves. Further out were the body boarders, and beyond them the real surfers, paddling their boards into the optimum position for catching the best waves to enable them to ride the breakers to shore.

Waves and revellers combined to create a constant background murmur, periodically pierced by the excited shrieks and squeals of children having a great time scaring themselves by venturing slightly further out, and into slightly bigger waves than they really should have. On the beach stood a row of worried parents, scanning the sea for the big wave that their children were fearfully praying for, then straining to see their precious little darling bob up through the surf as the wave rushed past. High on their observation scaffold towers sat the lifeguards, apparently relaxed, but clearly keeping a very close eye on proceedings. Occasionally, one blew a whistle or waved a flag, then signalled a swimmer to move within the safe zone designated by large flags which were flapping and tugging at their poles even though there was only a very light breeze.

Behind this shoreline activity, there was plenty of space for all those wishing to build sand castles or play beach cricket, football, Frisbee or bowls. Yet more people just relaxed, sun bathing, reading, chatting, or dozing the time away.

Eventually, our gaze reached the sand dunes running along the back of the beach, ranging up to fifty or sixty feet high and hiding various holiday camps, camping grounds and a golf course. Here and there were the remains of ancient churches, old mine shafts, and disused quarries. At the far end stood higher cliffs protecting the town of Perranporth, with rock pools at their foot, full of children trying to catch crabs. Before this, a stream flowed into the sea, with scores of children trying to dam it, build bridges or steppingstones, or just splashing around. In short, great fun was being had by all, including us.

In the mid afternoon, early August baking sun, burdened down by full packs, we marched straight up the middle of the beach, no doubt looking like two ants carrying a prize back to the nest. Sure, we were boiling alive, and received plenty of strange, but sympathetic looks from those dressed more suitably and relaxing in the sun, but we were in our element. We both love hot weather, and believe there is nothing better than the sun beating on your back for making you feel good. We were now making rapid progress, it was a beautiful day in an idyllic setting, and we had three weeks of the same to look forward to. We were surrounded by people enjoying themselves, particularly when they saw us.

We started to feel euphoric, and congratulated ourselves on our careful planning, which had timed our walk along the beach to coincide with the tide being well out, thereby enabling us to avoid dune walking, which can be very tiring if the sand is soft and the dunes hilly with a wandering path. When starting a walk like this, careful planning of the start time and use of a guidebook is recommended, as is retention of all maps.

For the whole two miles along the beach, we could see our target: the cliffs at Perranporth. As always in situations like that, initially the cliffs seemed to be retreating in front of us, and after twenty minutes I became rather discouraged, feeling progress to be extremely poor. I turned round to see how far we had come, to discover that we were in fact at least half way across the bay. This psychological boost immediately revitalised me and the cliffs started to grow larger, gain detail and appear closer, and the closer we got, the faster they seemed to grow and near. Soon we noticed the steps leading from the rocks up to the cliffs, and realised our next task was to climb them: 'That don't impress me much'.

That phrase is the title of a song by Shania Twain, which was popular at the time of our walk. From here on, every time we came to a tough section, made a pig's ear of something, the weather turned bad, or we just felt like it, one of us would start to sing the song, words adapted to the particular situation:

OK, so you're a big cliff. That don't impress me much.
So you've got the height but have you got the touch.
Don't get me wrong, I think you're all right,
But that won't keep me warm in the middle of the night.
OoooOOoohh Wah, wah, wah

Mad, I know, but it had such a bouncy rhythm and great words for stupid variations that it became impossible to sing without feeling good.

It was probably here that our unwritten challenge / rule was invented: never stop on an up. At the time, these steps seemed quite significant, but we both continued our march across the flat sand and straight up the steps without a pause. Nothing had been said, but the gauntlet had been thrown: first one to stop on an up would lose a very serious psychological point. Not that we were competitive or anything.

Having climbed the steps to the cliff top road, Phil went into Perranporth to buy new batteries for his camera and fill our water bottles, while I fell asleep on a bench carefully looking after our gear. When Phil returned we decided to take stock of our situation. We had now covered eleven official miles, but Phil had actually done seventeen, and me nearer nineteen. Our day one minimum target was fifteen official miles, but we had also thought of nineteen as an optimistic maximum.

Any experienced walker or expedition leader will tell you it is very important not to overdo things on your first day. You should stop while you feel good, and give your body a chance to acclimatise to the additional stresses you are putting on it. If you feel great the next day, you can push a bit harder then.

We consulted the map and realised that if we stopped at our minimum target we would be on the near side of a steep descent, with the only accessible pubs on the other side, which would mean down-up to pub, down-up from pub, and down-up again in the morning. (What do you mean, we didn't *have* to go to the pub? What kind of talk is that?)

Phil suddenly also realised that, with the tide out as it was now, but might not be the next morning, we could walk round the next stretch of headland on the beach, thereby avoiding several ups and downs. That clinched it: 'It's only five o'clock, we both feel good, let's go for it.' Our first mistake.

Had we looked in the guide, we would have seen that it suggested the beach walk possibility, and listed tide times, as we had discovered after making a mess of crossing The Gannel river. In any case, we should have been able to work them out roughly enough for our purposes. We weren't about to waste time planning or thinking, or reading a guide book of the route just because we were carrying it with us and had already made one big avoidable error that day due to not reading it.

Having scrambled our way on slippery rocks round to Trevaunance Cove, we stopped for a pint in Trevaunance Point Hotel so we would know what we were missing that evening. Phil had persuaded me that Guinness was the best pint for me to drink if I was looking to lose weight but wanted plenty of energy for the walking. 'It doesn't put any weight on and is full of iron. When did you last see a fat Irish navvy?'

Phil being as fat as an anorexic tapeworm and wiry as a cheap coat hanger, I thought he might know, so spent most of the next three weeks doing my bit to support the Irish economy. We sat in the beer garden, it began to get chilly and we stiffened up a bit, but didn't worry since we only had four miles to go. Ominously, Phil mentioned that he had felt a couple of twinges from some small blisters forming on his feet. If I had known then what I know now…

We set off round Newdowns Head, soon hearing music in front of us, which steadily grew louder. After a mile we passed the providers – a group of New Age travellers who had set up camp in what they thought would be a great location to view the eclipse.

There was a mess of old buses, coaches, cars and very battered caravans lining the cliff top and several dogs were scavenging around, two of which came onto the path and tried to scare us off by barking and snarling at us, and might have succeeded if we hadn't been so knackered and desperate to end what was rapidly becoming a painful addition to our day. Phil suggested I ask someone if we could spend the night with them, but I assumed he was joking and pressed on. We only had just over a mile to go but were feeling really tired now, Phil complaining more frequently about his blisters, and me beginning to suffer the first twinges of 'nappy rash', caused by my trousers rubbing the top of my legs.

We heard large machinery in action, and soon came across several police vehicles and diggers, placing boulders across all entrances to fields leading to the headlands.

We crawled up the final stretch of road, finding it very hard going, but eventually arrived at the site, where reception was closed, it being nearly 8.00p.m. I went in search of authority while Phil collapsed in a heap just managing to summon up enough energy to tell everyone he had walked at least twenty-five miles that day…now he had these bad blisters…he was doing the South West Coastal Path…twenty-seven miles is too far in one day…he had wanted to stop earlier but his mate had insisted on twenty-eight miles…it's a much tougher walk than he had expected…he would keep going, but his mate was a sadist…how could he be expected to carry his heavy pack up and down hills for thirty miles a day…nearly eaten by some wild dogs…probably more like thirty-two…

I found the owner and gained permission to erect the tent, which I did while Phil continued to ensure all campers were fully briefed on his physical condition and unreasonable companion. We changed as quickly as we could, and I tried to ring *the* St Agnes taxi, but there was no answer. It was 9.40p.m., dark, with a fairly thick mist swirling around before we were to be seen keeping ourselves upright by leaning on the front gate, pathetically praying for a quick lift into St. Agnes before everywhere closed. Walking one mile into town was out of the question.

I started to tell Phil how ridiculous we were being. Everyone knows that in camping, any time after 8.00p.m. is late, since campers operate on daylight hours. Anyone going into town would have left hours earlier, and most were at that moment tucked up in bed with a good hot meal and several pints inside them. We weren't in the right rhythm, being still on London time. We should have stopped at the fifteen-mile mark as I had suggested.

Phil has that great ability to totally ignore other people's whinging, probably because he is so used to other people ignoring his. He was stupidly looking up the road hoping for a lift from a passing car. Suddenly he grabbed my arm excitedly: 'Look'.

Incredibly, he had spotted the illuminated taxi sign coming along the road through the mist. It slowed and turned in at the gate, bringing the rest of the taxi with it. Some other campers had finished their evening's activities and grabbed a lift back to the site. I have a certain reputation for being lucky, but this was excessive even for me – that fox I had seen when walking to the station was really doing its stuff.

The driver agreed to drop us at one of the pubs in town and return at 11.15p.m. to take us back to the site. We knew we had no hope of getting a hot meal, but perhaps we could persuade the bar staff in the pub to rustle up a sandwich or something.

'No chance' said the barmaid, 'we finish all food at 9.00p.m.'

I turned to give Phil a 'that don't impress me much' look, but was stopped in mid grimace by the barmaid:

'Why don't you buy some fish and chips across the road? They don't close for another five minutes.'

Fish and chips for the second night running, but I was certainly not complaining, particularly as they cooked it fresh, to perfection, and in massive portions. We sat on a bench outside the pub, washing our fish down with pints of Guinness while congratulating ourselves on a brilliant start to the walk. I love it when a lot of detailed planning slots neatly into place, but it's far more satisfying when a complete muddle suddenly comes good! To confirm our marvellous luck, just as I was crumpling up my paper with a satisfied grin on my face, it started to rain, but we were able to move inside to join some locals at the bar for a final pint before our lift appeared exactly on schedule.

Having arrived so late the previous night, we clearly had no idea what our fellow campers were like, but it was quite a shock in the morning to find ourselves surrounded by people speaking Dutch. On thinking about it, we realised that the Dutch probably like places like Cornwall due to the hills and cliffs, which are sadly lacking in The Netherlands. There is also the fact that Dutch people seem to like the English and can speak English better than like what we does.

As soon as I went to put them on, I discovered the awful truth: my boots were on their last legs. There was a two-inch tear along the seam between sole and upper running next to the little toe of my left boot. I couldn't complain, having, as detailed in the Kit List, worn them almost permanently for the past three years without a single twinge or blister. Sorry, Phil, didn't mean to say blister, I promise not to mention blisters again. What's that on your foot?

It had rained in the night, so we decided to take our time getting ready while trying to dry the tent. We soon had bits of it lying on every blade of grass where the sun was shining, much of the site being shaded by trees. I went off for a shower, to find that they were prison camp style, the water coming out of a series of short pipes connected to one long straight one running the length of the ceiling. Who cares? They were hot and wet.

I returned to find Phil drinking a cup of coffee. He explained how he had used subtlety and charm to cadge it off our neighbours: 'I waited until their kettle was boiling, then said "I could murder a coffee, but we haven't got any..."'

I was so ashamed of his audacity I could hardly drink.

This being only our second day, and with many miles to cover, we should have been keen to get going, but in practice were looking for excuses to delay and didn't leave until well after eleven. We both found it, initially, very tough as we were still stiff from the previous, stupidly long, day, but I was soon into my stride, and Phil was soon into his blisters. We decided to use our walking sticks, having carried them in our rucksacks all the previous day, and it only took a few minutes to explain to Phil how they worked.

Unreasonably, as Phil had cunningly predicted it might be the day before, the tide was now in, so we couldn't beach walk. Nevertheless, there were only a couple of ups and downs around Sally's Bottom and we made fair progress. You may recall that I was due to return to Essex for my niece Sally's twenty-first birthday party at the weekend, and I toyed with making an announcement there: 'I had a great time up and down around Sally's Bottom.'

That kind of joke sometimes falls flat on young lady's parents, and she has a rough, tough and jealous boyfriend, so I decided to adopt my customary tactful approach and just whisper it to some of the guys.

There was an old disused airfield near Sally's Bottom, which was flat and smooth.

We shot along, although Phil was mentioning his blisters more frequently. It was raining quite heavily as we came down the hill into Portreath, so we were pleased to see Portreath Arms Hotel was open. Phil suggested we should have a kitty so that all joint expenses, like food and drink, could be bought with no worries about who was paying most. As mentioned previously, Phil had loads of money while I had hardly any; he now agreed to sub me my contribution to the kitty, and let me hold it. Throughout the rest of the walk, he was constantly asking whether there was enough money left, and was I sure I wasn't spending my own money by mistake. Now, that's what I call

~~stupid~~ friendship. I managed the situation by paying for everything I could with my credit cards, and using Phil's cash for the rest. This worked out very well, and my cash strapped embarrassment was over.

We had a good laugh in the pub. The landlord was originally from West London, so knew a lot of Phil's favourite haunts, and a group of locals made us most welcome. The ceiling was covered with baseball caps, which pleased Phil no end since he is a big baseball fan. He bet the owner that he could state one he didn't have: Atlanta Braves. This proved to be correct, Phil explaining that they are his favourite team and, for some unknown reason, despite being one of the best and most popular teams, their cap is very rarely found anywhere other than Atlanta. He then gave a fascinating potted history of the club's major achievements – most successful team of the nineties in the national league, and overall second only to the New York Yankees of the American League. Since I thought a pitcher was something to put water in, I was forced to conclude that when it comes to useless information, Phil is a font.

Given the miles still ahead of us, and the macho image we were attempting to project, we were not able to sit out the rain, so put on our wet gear, which caused much merriment, and hit the door, literally in Phil's case. Just for a change, there was a steep hill out of town: we were already noticing how every time we had a nice break for refreshment, we paid for it with a steep climb. We were also beginning to realise that the time and effort involved in this walk would be far greater than suggested by the miles covered. It would be how well we coped with the ups and downs, which determined our success. Fortunately, the next few miles were mainly flat, with few descents into inlets, so we made good progress, and were rewarded by the sight of an ice cream van parked in a sightseeing area.

We just had to have a Cornish ice cream – we were in Cornwall, it tastes superb, and the van owner stocked nothing else. On a bench made from a tree trunk, dangling our legs over the cliff like two school kids on an outing and licking our cones, we congratulated ourselves on finishing map one, or, if you want to be pedantic, map two. With a flourish, Phil produced the next map, which would cover our journey all round Land's End and back east along the south coast as far as Praa Sands 'provided you don't leave it somewhere, that is.'

On setting off again, the route went out and round Godrevy Point. We decided, however, that we had seen enough headlands for that day, and would cut a mile from the route by keeping to the road. This proved to be a very bad decision as within half a mile it became extremely tough on our feet. By now, Phil's blisters were really giving him gyp, as he constantly explained:

'I'm really fit, the ups and downs are no problem; it's these blisters – you don't know the pain I'm in.'

As you may recall, Phil had suffered from glaucoma for many years and used eye drops to control it, with side effects including slowing his heart beat, which in turn tended to slow him down on tough sections, and was one of the reasons why I was carrying the tent. Once we had started the walk, and the tent was safely stowed in my rucksack, Phil had revealed that he was now on a new prescription, Xalatan, which did not have side effects, so he was now much fitter. This was certainly true, because I knew how much I was struggling up some of the hills, yet Phil always stuck right behind me and never seemed out of breath at the top, grinning at me like some gargoyle or a devil on my back. I began to suspect that he had hooked a rope to my rucksack and was using me to pull him up.

If the truth were known it was probably only Phil's blisters that were stopping him leaving me for dust, not that there was a competitive edge to it. Luckily, his new eye drops were no good at toughening feet. You might ask whether there should have been a redistribution of weight, with Phil perhaps carrying most to compensate for my relative portliness, but I had already revealed my desire to lose weight, so he countered any such suggestion by claiming the extra weight was helping me achieve my goal. Ever thoughtful of his fellow man.

Walking on the road was really bad news, but having come that far, there was nothing for it but to continue pounding along, focusing on our next objective, Gwithian Bridge. Soon it came in sight, and I told Phil he could have a rest there. I knew I risked being too soft on him, but didn't want him to crack. Yet.

It seemed to take forever to reach the bridge, but eventually we threw ourselves down next to it, pulled out our water bottles and agreed to minimise road walking from then on. It was only half a mile to our intended campsite, but all on tarmac – just so painful. Despite my telling them to stop complaining, my feet throbbed from the constant pounding on the hard surface. I knew they would recover as soon as we were back on more yielding ground, but could see Phil's were in much worse shape.

At last, we turned into the site and booked in at the caravan, which served as an office. We dumped our rucksacks outside while signing in, and left them there while Miss Campsite showed us the two pitches we could select from, both within fifty yards of the van, so we were back within a couple of minutes. We picked up our bags, returned to the chosen spot, and I started to get the tent out. Suddenly Phil realised that his sleeping bag and ground mat were missing, and immediately decided they must have been stolen while we were inspecting the sites and our bags were unattended outside the office van. While he went to ask Miss Campsite if anything had been handed in, I had a look around to see if they had rolled away or fallen nearby.

Phil returned and we shared our disappointment. I was very sceptical of the idea that they had been stolen, thinking it was far more likely they had fallen out somewhere on the day's walk. Phil might not have felt them go, and since I was usually in front, I would not have seen. Phil insisted he had had them both securely held in place under the top flap of his rucksack, so they could not have fallen out. I pointed out that this meant in order to steal them, a thief would have had to undo the two securing straps, take out the bag and mat, then do the straps up again: we had not been away that long.

Phil was still convinced that they had been stolen, but decided to walk back the half mile to Gwithian Bridge, our last resting place, just in case they had come out there. While he was gone I erected the tent, had another look round and asked some other campers if they had seen anything: no luck.

Phil returned empty handed, but philosophical. 'You leave it for two minutes, and some bastard nicks your tent. Still, no good worrying about it.'

He kept saying tent instead of sleeping bag / ground mat. I was mug enough to comment on his stoicism, receiving the response; 'When you've had as tough a life as me, you get used to being kicked in the teeth. If I let it get to me, I would never do anything. I don't expect much from life, so I'm not disappointed when I don't get much.'

We spent the evening in a very local pub. I don't know whether it is an old Cornish custom for pubs to have their ceilings lined with a collection of some favourite item. Portreath's had specialised in baseball caps, and this one had row after row of mugs. Most were fairly standard, but some were ornate, hand painted beauties.

The barmaid was extremely friendly and lovely, particularly as she served me next every time I went to the bar, and 'forced' the kitchen to squeeze us in although we arrived late and they were already fully booked. At the end of her shift, she came over and started chatting to us. She lived in a local house, which overlooked the sea, loved it, and was very jealous of our trip, sympathising with Phil over the theft of his tent. When she asked whether mine was big enough for two, she was temporarily confused when I replied there was no way I would let him near my sleeping bag, but if she was offering to share... When Phil started to tell her all about his blisters, she suddenly remembered an important engagement and left.

There seemed little point in hanging around, so we headed back. Phil informed me:

'I thought you were in there until she shot off like that.'

'So did I. Good of you to bring up your bleeding feet, you great wazzock.'

I was not really so convinced of my potential prospects, feeling she had given every indication of being very happily married. In any case, I chat up anything in a skirt, including the occasional Scot, but run a mile if I receive a positive response – all mouth and no trousers, as Lynn says.

Before falling asleep, I mentally reviewed the trip so far. We had covered thirty-three official miles in two days, leaving only forty-seven to go in three whole days with Saturday morning in reserve if needed, my train back to London not leaving until nearly midday. Should be a piece of cake provided we kept roughly to the route and cut down on the detours.

Phil's blinking blisters were the only real threat to a successful first week, but I thought they would probably start to improve as his feet toughened up. Wimp! I couldn't really understand how he had got them: he did a tremendous amount of walking, and was wearing old, worn-in boots.

I felt fine, the throbbing in my feet having gone, and the minor threat of blisters from my sandals having disappeared as soon as I had put my wonderful Approach boots on. I was also very snug and comfortable in my sleeping bag with ground mat underneath, made even more pleasurable by the fact that Phil was having to make do with putting on his spare clothing. I could have let him borrow my mat, but you know how it is. Anyway, he obviously needed toughening up, and shouldn't have put his foot in it for me with the barmaid.

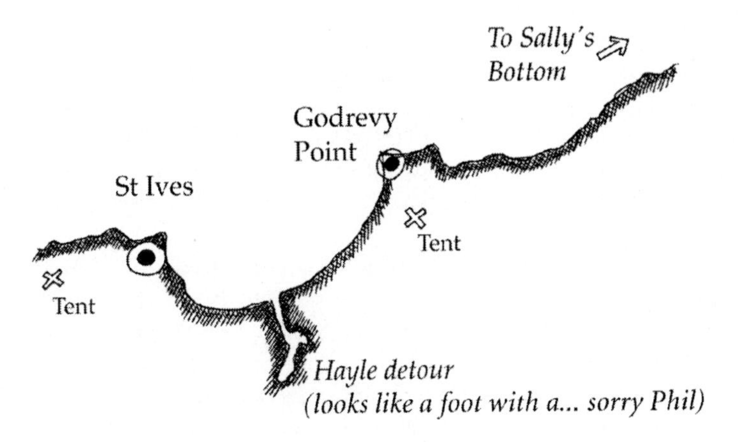
To Sally's Bottom
Godrevy Point
St Ives
Tent
Tent
Hayle detour
(looks like a foot with a... sorry Phil)

Mister Blister Meets Sister Blister

It was raining heavily when I woke next morning. I am a great riser, except when cold or wet, so lay back for a while hoping it would stop, listening to the rain pattering on the canvas, and betting with myself which drop would run down the flysheet soonest. Fairly soon I became bored and decided to get up and visit the office van to check whether Phil's gear had been handed in (no), and try to obtain a weather forecast: rain all day, then dry all night.

Phil spent ages trying to sort out his feet with a ridiculous number of plasters. Eventually, I decided to cheer him up: 'Come on, Mister Blister, I'm sure your feet will settle down after a mile or two.'

We donned our waterproofs, packed the soaking wet tent, which made it twice as heavy, and left by the back entrance straight onto the sand dunes framing St Ives Bay. Unfortunately, we had not read the guide's tide table or worked it out for ourselves, so the tide being too high for us to walk on the beach round Strap Rocks came as a surprise. We started to follow the dune path towards Black Cliff and the Towans, two miles along the bay.

The dunes were bad news, with soft wet sand that clung to our boots, making them very heavy, and we sank a couple of inches into it every step. Rather than being a constant height with a ridge we could walk along, the dunes consisted of a seemingly never ending rolling series of ten to fifteen-foot high lateral dunes, which we had to keep dragging ourselves up and down. Marker poles were placed fairly regularly indicating the route of the path, but it seemed to meander along with no obvious logic: it was not straight, and did not keep to a constant height contour or relatively solid sand.

As soon as we were clear of the outcrops around Strap Rocks and Peter's Point, which made me think of my son Peter, who would have given anything not to be there, we searched desperately for a way down to the beach. For about a hundred yards we had no luck, there being a drop of a good fifty feet along the front edge of the dunes, but then came across a scramble path down.

As the tide receded, it revealed another broad, flat, beautiful sandy beach. Too late, we realised how much better it would have been for us to spend an extra hour or so at the camp site, possibly drying the tent and really sorting Phil's blisters, then once the tide had gone out enough, set off along the beach. Now, we were hardly any further along than we would have been, but were considerably more knackered due to scrambling around in the dunes, and still had a heavy wet tent (me) and scary wet blisters (Phil).

We intended to spend that night somewhere around St Ives, which was temptingly close, only four miles round the bay, the pain being that it was twelve miles walk away due to the major detour inland required to negotiate the river Hayle.

On the beach the sand was still rather soft, but nowhere near as bad as on the dunes. Being broad and flat, it made for rapid progress and little chance of getting lost. Walking in such an environment, particularly in wind and rain, is incredibly invigorating and exhilarating, blowing away any cobwebs from the mind and forcing one to forget any worries.

Consulting the map told us there were several camp sites behind the dunes, and the towns of Hayle and St Ives were close by, yet the beach was deserted, as had been most of the others we had come across, many with good surfing waves and plenty of accommodation close by. Okay, it was raining now, but at other places the sun had been baking down. Where was everybody? The eclipse was only a week away.

One of the few people we did see was swimming fifty yards off shore. As we walked along, he slowly progressed towards, then past us, carrying on, and on, and on, swimming the full length of the bay. The things some people do.

As we neared Black Cliff and Hayle Bar, we became increasingly frustrated – we could see St Ives two miles away but had to go on a stupid eight-mile detour. Much of this would be on busy roads (danger, fumes and tarmac, Phil's favourite combination) passing mud flats, which are not particularly attractive at low tide, and in the hissing rain. The exhilaration of our beach walk soon evaporated.

Our detour started with a climb up off the beach onto Black Cliff and on to Hayle harbour via the Towans, estates of holiday homes and flats, each estate named something Towans, including Mexico Towans and Riviere Towans, presumably in the vain hope that it would convince people how good the climate is.

It was still raining, and Phil was still suffering from his blisters, and ensuring I knew it. I had prepared jokes along the lines of 'Hayle fellow, well met' and 'Make Hayle while the sun shines', now replaced by 'F…ing Hayle' and 'at least it's rain, not Hayle,' but even these did not lift Phil's spirits.

As we entered town, we started to smell freshly baked bread, cakes and pastry, so followed our noses to Philp's Bakery where we stopped for tea and a Cornish pasty, taking full advantage of some sheltered seats outside the shop. I soon had that smug tummy feeling, but Phil was still not a happy bunny until a young couple limped round the corner, whereupon he and the woman recognised each other as fellow sufferers and their eyes lit up. Phil launched into his blister routine, and the three of them rapidly became deep in discussion. After a few minutes, however, the man lost interest and went to get some tea, leaving his companion and Phil well on the way towards a PhD on the finer points of blistology: shape, overall surface area, depth (average and maximum), layers of skin above and below, consistency and colours of the disgusting fluids inside, level of pain and under what circumstances it worsened (both agreed tarmac should be banned). Sale of custard tarts took a temporary dive.

Eventually, the topic of conversation changed – to how to treat blisters. Mrs Limp was a world-renowned expert and had three sections of her rucksack devoted to various potions, lotions, needles, swabs, pads and plasters. Mister Blister had met Sister Blister! She started by giving Phil a box of plasters and bottle of surgical spirit to toughen his feet. Several pieces of Mole Skin and Compeed followed, which she explained had nothing to do with cruelty to small furry animals, but were used to ease the pain of existing blisters and help prevent new ones developing. I don't think her partner was particularly impressed at her generosity.

Having spent the best part of an hour discussing blisters and sorting Phil out, thirty seconds were spent by Sister Blister telling us that they were on honeymoon, doing part of the coastal path, this was their third day, and they had only covered two miles that morning so far. They hobbled off for their first argument – 'Fancy giving away half of our wedding presents like that' – intending to stop the night in St Ives. Only another five miles – pathetic: we would obviously soon be overtaking them. We actually never saw them again, so if they happen to be reading this, all the best from me, and blisterly love from Phil.

Phil couldn't wait to test his new remedies, so continued the slump in the custard tart market by peeling off the several layers of plasters already on his feet, carefully inspecting his blisters armed with his new technical knowledge, and whacking on a silly amount of Moleskin, all the time giving a running commentary to anyone unfortunate enough to be in range. When we finally got moving, I realised I was starting to get 'nappy rash' again, which I knew from bitter experience to be particularly vicious in wet conditions due to repeated rubbing of wet material against the raw flesh of the inner thigh.

Two miles of road walking followed, much on grass verges, but some very dangerous sections on the B3301, where we were forced to walk in the busy road with high hedges helping to obscure drivers' view of us. The guide suggested a short cut to avoid a particularly bad section, and we decided to go for it. This was our first mistake as, at Lelant Saltings station, the suggested route did not seem to exist, but three other possibilities did.

While we were pondering our predicament, a tall, slim man in well used walking gear, with a reasonably sized rucksack, large floppy felt hat, and a very slow, measured gait, came to join us: 'Hi, I'm doing the coastal path and trying to follow a short cut, but seem to have lost it'.

He informed us he was doing the whole six hundred miles from Minehead to Bournemouth, so far averaging twelve miles a day. Right in line with our original idea – but he was actually doing it. Born locally, he had not been back for twenty years, but, having recently been made redundant, he was taking the opportunity to get back to his roots.

One of my failings is that I just can't resist making some inane comment in the middle of the most serious conversation, so out it popped before I could stop it: 'If you weren't on this walk, you wouldn't have lost your route, so you wouldn't need to get back to your roots'. Phil started to inspect his blisters. The man was obviously a loner, because he asked which way we were going, and said he would try one of the others, bye. Fancy giving up the opportunity of a few hours, or even days, in our company.

We carried on through some playing fields, and found a path at the other end leading the way we wanted, so thought we must have guessed right. After two hundred yards, however, this path died amongst a mess of brambles, thistles and other spiky plants, which did not particularly amuse me as, the sun having finally shown its face, I had just taken the legs off of my combi-shorts. We decided to fight our way over a stream and the rail track onto the mud flats, the tide now being well out. Getting there was much easier than we expected, but the mud was unbelievably slippery and slimy, so that any step could be our last. We made our way extremely gingerly, with arms waving to keep balance, to Lelant Station and gratefully placed our feet back on a stable surface. It had taken us an hour and a half to do a short cut of one third of a mile.

We crossed the rail track and went inland uphill to St Uny's church, which we discovered to be the start of St Michael's Way. In medieval times, pilgrims travelled from all over Europe to the cathedral at Santiago de Compostela in Spain. St Michael had apparently started his pilgrimage at this church, made his way across the Cornish peninsular to the rocky island now known as St Michael's Mount in Mount's Bay near Penzance and thence to Spain.

Feeling that, so far that day we had made a pig's purse out of a silk ear, we were very pleased to make our way through a golf course down to Porth Kidney Sands where we were able to enjoy a nice beach walk for a very pleasant mile. We tried to resist looking over our shoulders at Black Cliff, only half a mile away in a straight line, but an awful six-mile walk behind us.

Very soon we approached Carrack Gladden rocks, where I was met by an extremely beautiful, tall, slim young lady with long flowing nut-brown hair, making her lithesome way incredibly gracefully down the rocks and towards the shore. I couldn't resist: 'Wow, my first mermaid!' She gave me such a wonderful smile my legs almost gave way and I half expected her to dive into the water and a fish tail to appear as she swam away. The pain of Hayle was forgotten.

Carrack Gladden rocks can be walked round at absolute low tide, but it was still at least an hour before that, and the rocks still had a good two feet of water covering their seaward edge. I was all for wading round, but Phil would have none of it: 'my plasters might react to the salt water' so we made our way up the rocks – not quite as pleasant a sight as the mermaid had been coming down – and along an amazingly overgrown path for a couple of hundred yards before returning to the beach on Carbis Bay. There were only three people at that end of the beach, a toddler running around playing sand castles and splashing in puddles, and two adults tangled together in a very interesting position. As we came onto the beach the man gave me a filthy look, as if I was a peeping tom or something, and I looked hurriedly away.

There was a cafe in the middle of the bay, and we decided to stop for an ice cream. We sat on the beach wall and watched what seemed to be a mother's club outing: there were a dozen young children having much fun, while seven or eight women sat around chatting. Suddenly I noticed an extremely good quality sand sculpture of Neptune riding on a Dolphin, but from my angle it looked as though he could have been doing more than just riding it. Coupled with my experience when first stepping onto the beach, I was beginning to wonder if there was something in the water at Carbis Bay, or I was missing Lynn more than I realised. Perhaps I really had met a mermaid and she had cast some evil spell on me – I wondered what her next move might be, and almost fell off the wall.

Having finished our ices, we were just setting off again when we were approached by two lovely Dutch girls who wanted to look at our maps. Phil happened to mention to them that he had some blisters, and they showed immense sympathy, as had the woman in Hayle. I wondered whether I should get myself some blisters – I could hardly tell them about my nappy rash…could I?

They were finding the walking too hard, and intended to hitch a lift into Hayle and hire bikes. I said 'you've got no chance hitching a lift. Who would stop for two gorgeous girls like you?' and received a very friendly smile in reply. Smooth or what? Much better than 'well, Hayle a taxi, then' and infinitely superior to any wise cracks involving local bikes, which would no doubt have resulted in a black eye.

Incredibly, Phil was able to tell them exactly where the bike hire shop was, how much they charged, and what time it closed. He pointed out later that we had walked past it a couple of hours earlier, but there was no way I would have noticed or remembered anything useful like that.

They asked where we were staying, and when we said we were flexible, suggested we stopped at their camp site and they would meet us there later: 'It's a couple of miles the other side of St Ives, lovely, with great showers, facilities and a shop. I'm Ingrid and this is Birgide.' Coincidentally, we had been thinking of stopping at that very site, honest – even Phil's blisters suddenly felt better, providing he stood still.

Waving goodbye, we set off up the obligatory hill, Phil suffering very badly – even the thought of spending an evening with two lovely ladies was not keeping the pain at bay. He had developed stock phrases which he kept repeating randomly ad nauseum, sometimes the same one twice in a row, or muddling them together:

'…you don't know the pain I'm in…it's agony…it's not the fitness…it's alright for you…I'm as fit as anything…you don't get pain…I can easily cope with the ups and downs…I'm as fit as anything…it's these blisters…you don't get the pain I'm in, blisters every step it's fitness agony…up fits of pain down…all for you right you're in right agony…it's a fit up…I'm bliss fit no pain…you pain fits…know don't fit the pain again…I'm in right any pain thing…I'm a right pain…' whine, whine, moan, moan.

He might have been in some pain, I guess, but not as much as me having to listen to him whining on about his feet. To try to make him realise he was not the only one in pain, I foolishly mentioned my nappy rash, although it wasn't too bad at the time. Phil was delighted and cheered up immediately:

'Ha ha ha. That's terrible, can be worse than blisters. I feel sorry for you. BR can be a right killer.'

'What do you mean BR?'

'Bollock Rub. Revenge of the Woman Walker.'

He had certainly hit the spot.

Although I was keen to press on and get the day's walking over, I agreed to a stop at Porthminster cafe for a pot of tea. The location was, again, magnificent, with seats on the terrace, which formed part of the sea wall, giving fantastic views of the beaches and cliffs all along the coast. We were easily able to see where we had started that morning – it was only three miles away. Further excessive generosity on my part followed when I agreed to let Phil buy a new sleeping bag for himself.

St Ives, late in the afternoon in mid August, was heaving with humanity, just the place to wander through with a large rucksack. At least Phil couldn't clout people round the ear with his mat since he'd lost it, but that didn't stop him poking them with his stick. In every shop that looked as though it could possibly sell camping gear, Phil explained his predicament with the opening gambit 'Two minutes and someone nicks me tent', thereby confusing the shop staff who could see I was carrying one big enough for four people.

Once I translated for him, we received the same reply in each of the first three shops: 'You won't be able to buy a sleeping bag in St Ives' stated as though it was a ridiculous question: why would any one want such a thing? I began to get bored – I had a lovely warm sleeping bag – and started to look at silly cards in shop windows, when I received a dig in the ribs: 'John. I've found one.'

I followed Phil round the corner and into a hardware store. 'We've only been doing them for six months, and we've sold loads. Don't know why we didn't think of it before. We might even put a card in the window advertising them soon'. And they say the entrepreneurial spirit is dead.

Although it was tempting to stop for a couple of beers and a look round, we decided to push on to the camp site, settle in, eat, see whether the Dutchesses were around, and when they weren't, get a taxi back to St Ives for the evening. The site, Trevalgan Holiday farm, was two miles out, along roads, and uphill for the first mile and a half. This combination at the end of a long hard day soon had Phil doing one of the best imitations of 'old man Steptoe with backache and in-growing toe nails' I have ever seen. He was virtually crawling by the time we crossed two fields and came to a very easy stile leading onto the site: 'Have I got to get over that?' as if it was an army fitness course wall.

I booked in, found our site and put the tent up, while Phil trailed around telling me for the seventy-third to seventy-sixth times that I didn't know the pain he was in, every step was agony, it's the blisters, he is as fit as anything, it was alright for me, I didn't get pain… did I keep going on about my BR?

The campsite lived up to the girls' description. Reception had a general purpose shop and take away fish and chip shop attached, with several sets of tables and chairs on a paved area in front, and the fine shower block included a large, spotlessly clean changing area, a laundry and drying room. Our site was, thankfully, quite close to reception, on what would normally have been the sports field – we were in one of the goal mouths of the football pitch – but had been commandeered for extra camping spaces due to the anticipated heavy demand surrounding the eclipse.

Our evening meal consisted of a cup of tea and, for the fourth night running, fish and chips. By the time we had finished, Birgide and Ingrid had arrived, parked their bikes, bought two beers in the camp shop and come to join us. Excellent. I began to look forward to a nice relaxing evening, chatting to two beautiful women and slowly getting drunk.

'Okay, John, go and phone a taxi. We've got to see St Ives.'

'But Phil…'

'You'll love it; it's a top place – one of the highlights of the walk.'

Phil tends to think wherever you are going next is 'a top place', but I must confess his enthusiasm is contagious. Anyway, I knew it was pointless arguing since he would have just kept on about it until I finally gave in, and always had the blister story up his sleeve, and in full colour on his feet, if he wanted to see a double Dutch disappearing trick.

While we waited for the taxi, I tried to persuade the girls to come with us, Phil supporting my efforts by telling them how some nice person had nicked his tent and I hadn't even lent him my mat. Hearing the slight edge of disappointment in his voice made me feel guilty, and I guess it was a bit cruel. Still, I'm sure Phil didn't mean it.

I had already showered and changed and Phil was supposed to be doing so before the taxi arrived, but was far too busy extolling the virtues of the Scilly Isles, and explaining how it's a top place and they mustn't miss it, so it wasn't until after the taxi arrived that he limped off towards our tent to change.

You may recall Phil's sartorial elegance and natural clothes sense from my description of him at Reading station. Now, he was wearing a pair of badly worn walking boots, two pairs of disgusting, filthy and smelly socks, dirty and incredibly creased track suit bottoms, with one leg half tucked into a sock, an old T-shirt, a moth-eaten shell suit jacket, with his ever-present bumbag full of God knows what strapped round his waist, and LSE baseball cap on his head. He was unshaven and unwashed, although as far as I could tell he didn't actually stink. Despite his Steptoe shuffle, he was back within three minutes with no discernible change having occurred. The girls decided not to come with us.

In the taxi I reflected on success so far, and likelihood of reaching Penzance in time for my Saturday train: forty-eight miles after three days with thirty-two still to go. I thought we should make it quite comfortably if we did say thirteen miles on each of Thursday and Friday, leaving six miles for Saturday morning. I also realised it was just as well that the girls had decided not to come: they would probably have thrown themselves at me in a wild, sex frenzy, and I might not have had the energy to fight them off. Who knows what effect they may have had on my BR?

In St Ives, having booked the taxi for11.30p.m., we started to look around. After fifty yards Phil sprang his big surprise 'It's no good, I can't go on. You don't know the pa...' There is a limit to even my good nature, and given that Phil had insisted on going there, ruining a God-gifted evening with two lovely ladies and plenty of booze, this was hard to take.

'Okay. You sit there and talk to your blisters while I have a wander.' I toyed with dodging round the corner and grabbing a quick taxi back to camp for a final attempt to 'go Dutch', but just couldn't do it to him – though looking back I still cannot see why. Instead, I had a quick (subject to BR) mooch around, which revealed St Ives, as Phil had said, to be a top place. There was a good mix of old and new buildings and shops, with plenty of odd-shaped narrow, twisting back alleys to explore, a nice pub / restaurant area, harbour and front, including a small fleet of fishing boats and yachts. People were out in force, all having a great time, in high spirits, and as friendly as could be.

Within half an hour, I found myself back at Sloop Inn, the quay side pub where I had left Phil, and we spent the next two hours drinking, watching the yachts in the harbour while beautiful, scantily clad women walked past, and commenting on other customers, although I bet they had more fun commenting on us.

Being unattached, Phil loves to point out relationships that don't seem to be working, and, it must be said that there was a great example at a nearby table. Both looked desperately keen to think of something new to say to the other, but they also had an air of hostility about them as though they had recently been arguing. They didn't want to say anything that could give the other a chance to score points, so just sat there staring at their drinks in strained silence, while time crawled slowly by. Eventually they finished, but it was still too early to return to a frosty hotel room. To ease the tension, the man went for refills, was gone for ages, and when he returned brought with him an attractive woman he had been chatting to at the bar.

For some reason, that did not go down very well with his companion, but certainly did with Phil. Throughout the whole episode, he had been giving me a running commentary in his usual subtle megaphone whisper with accompanying gestures and pointing. When she downed her wine in one and stormed off, Phil was so excited he stamped his feet and let out a victory whoop – or was it a cry of pain?

Next morning, we decided to take advantage of the facilities: I would do some washing and treat my BR, while Phil sorted out his feet. My 'nutter' T-shirt attracted the attention of yet another Dutch couple, who proudly said: 'we saw your T-shirt. Our name is Nutter. We are a couple of Nutters.' I am still not completely sure that they understood the joke and were joining in – perhaps they thought I too was a Nutter, and not just a nutter. I offered to sell them it for two hundred pounds, but they said they might be Nutters, but they weren't stupid.

While I was taking the tent down, Phil saw the girls off for the last time, while filling his face with two sausage and egg butties, and a bacon one. It was midday when we left, but two when we reached the offshore rocks known as The Carracks, a mile and a half away. Can you guess the cause of our tardiness?

Every two-hundred yards, Phil had stopped to add a third pair of socks, swap Moleskin for Compeed, sob, tighten his boots, have a drink, try two pairs of socks, go back to normal plasters, wince, have a drink, loosen his boots, wail… while constantly reminding me that I didn't know the pain he was in, it wasn't the fitness, his stamina was great, but every step was agony. It was okay for me, I didn't have any blisters, I didn't have any pain, every step I took wasn't agony, but for him, it wasn't the ups and downs…

I enjoy schadenfreud as much as anyone, but this was OTT, and the constant whine was starting to annoy me. I was getting set to launch into an 'Oh, for God's sake, pull yourself together man' speech, pointing out that perhaps I wouldn't have minded a few ups and downs with Birgide, when two miracles occurred.

Firstly, I suddenly realised that we were supposed to be on holiday, and it didn't really matter if we didn't make it to Penzance that week. What really mattered was our friendship: useless wimp that he was, I should be more kind, considerate, understanding and sympathetic to my good friend Phil. Given the way I had gone straight back up Mount Kilimanjaro when I failed to reach the top first time, most people would assume there was little prospect of my giving up the Penzance target so easily. In fact, if you have really tuned into my psychology, you will realise that this was not a miracle at all, but right in line with my philosophy: Phil's blisters were not my fault, so if we failed because of them, there was no reason for me to be disappointed in myself.

Miracle number two (or one if you don't count the one above) was that we met someone in a worse state than Phil. A man and woman were on the walk but the man's knee was in a very bad way, and he was hardly able to put any weight on it. As had been the case with Sister Blister, Phil immediately struck up an affinity with this new fellow sufferer:

'Yeah, mate, I know how you feel. It's not the fitness, it's the constant pain...'

'Tell me about it! Every step. I'm fine on the up, fitness is great, it's the downs that get me.'

'She doesn't understand.'

'Course not, nor him.'

'She' and 'he' exchanged knowing glances; hoping things would improve now 'they' had met someone in a similar state. I considered revealing my BR to her, asking if there is a female equivalent, and if so, did she have it? Decided best not to.

The Carracks are famous for having gannets and seals on them, and that day was no exception, so we watched them gannet and seal for a short time, but Phil was keen to get moving. As soon as we had gone a few yards, Phil started to gloat. 'That poor bastard. He's had it. Did you see the state of his knee? He'll never make it. He wanted my stick, but no way. That stick has been what's kept me going. Best investment I ever made.'

This elation at another person's misfortune spurred Phil on at quite a reasonable pace for the next three miles, but unfortunately this consisted of the toughest part of the walk so far, with constant ups and downs and rough ground under foot. Phil slowly succumbed to the pain, and I felt I had to put him out of his misery: 'Phil, I've been thinking. We are supposed to be here for pleasure, and it is much tougher than either of us imagined. Why don't we just aim for Treen tonight, do a few more miles tomorrow, then pick up after the eclipse from wherever we get?'

I could see the tension, pain and sheer physical / mental concentration / willpower drain out of Phil's face. For a terrible moment I thought he was either going to cry or kiss me – I doubt if I could have coped with either. I hadn't realised the dramatic effect my offer would have on Phil, but saw immediately that, had I wanted to, now would have been the time to deliver the coup de grace 'Just joking. Of course we must stick to our target.'

At first Phil tried to pretend the thought of not walking all the way by Saturday had never entered his head, and it may be true that it hadn't. He thinks I am amazingly competitive, and would hate to admit failure having set myself a goal, so he was probably very loathe to make such a suggestion in case I agreed, then hated him forever. It was not long before he 'reluctantly'concurred, and his normal buoyant character started to reappear. Soon he was constantly repeating how we were meant to be on holiday, there was no point flogging on if we weren't enjoying it, we could enjoy the views much more at a slower pace... The old Phil was coming to the surface.

We continued the very tortuous route, taking more drink and viewing stops, and passing several groups of walkers, most of whom seemed to be Dutch, which was beginning to intrigue us. By the time we left the path and climbed the hill into Treen, Phil was in a bad way again. 'My feet are shot to pieces (a new phrase!), you don't know...' Help! Get me off this hill before I kill him.

The first farmhouse offering B&B was full, but the owner took one look at Phil and said 'I'll run up the road to see if my friend will take you,' which she did, and she did. We had an en suite double bedroom each, on either side of the staircase, in a wonderful old farmhouse. It was beautifully decorated throughout in very traditional and tasteful furnishings, the highlights being a marvellous old table in the dining room, and the superb original kitchen, with a full set of contemporary kitchen equipment, including witch's broomstick. Mrs Farmhouse was justifiably proud, and gave us a guided tour, pointing out the massive two hundred year old iron kettle and cauldron, which were still hanging in the beautifully preserved fireplace. I had a quick peep in the cauldron to check for eye of bat, foot of newt and tooth of toad, but was disappointed. To match the scale, several giant wood and metal stirrers, along with various tongs, hooks and pulleys completed the scene, and made it look more like a torture chamber than a modern kitchen. Speaking of torture, once we had dumped our bags, Phil decided to show me his blisters. Oh dear…

On the pad underneath each heel was an extremely good quality, classic blister, one and a half inches square, full of plasma and whatever else blisters are full of. Both were intact, and Phil kindly demonstrated how applying pressure to a point on one side caused the liquid inside to wodge around and well up on the opposite side. Most of his toes had at least one blister, and some had two or three vying for space, with a couple actually overlapping, presumably at different skin depths. There was a potential prize winning beauty stretching across the front and sides of one toe, very full with thin skin stretched tightly round looking fit to burst at any second.

All of these were eclipsed by the one on his right foot, on the pad just behind his toes. It had burst at the front, and both sides had split so that a piece of skin several layers thick and two inches square was hanging down like a hinged trapdoor. Underneath was an oozing mess of blood and gore with small pieces of gauze, plaster, moleskin and sock threads stuck in the gooey slime. Phil started to pull at some of these, and they seemed to be alive, clinging onto his weeping flesh and pulling bloody, squirming lumps of it up for closer inspection. A couple tugged free, small drops of god-knows what flicking onto the bedcover, and a fresh spurt of yellow and red ran across the half congealed battlefield beneath. Perhaps I had been a bit hard on him, telling everyone he was a whinging softy?

I pointed out two good things. Firstly, the pad of his left foot was fine, secondly, they were on his feet and not mine. He was not amused, and hobbled off to show Mrs Farmhouse, presumably hoping for more sympathy, which he duly received along with much advice. While he was putting this into practice, along with that received from Sister Blister, I had a long, hot, soothing shower and a much-deserved rest on my bed.

Treen's pub, The Gunnard's Head Hotel, was only a hundred yards away, at the end of the farm track, so we went for a couple of pints to test Phil's latest patches, back to the B&B so Phil could shower, then returned to the pub for our evening meal. We had a great time, there being more bar staff than customers, probably because ~~we were there~~ they had live music on Wednesdays and Fridays, so people went elsewhere on a Thursday if they could. Here the walls were decorated with rowing cups and paraphernalia, locals appearing prominently on the rolls of honour. We had our standard Guinness and fish, although this time we each had three small fish of different local varieties on one plate.

Hobbling back from the pub, Phil suddenly disappeared with a surprised groan into a deep trench, emerging covered in mud, moaning about some fool digging holes in the middle of the path. I suggested that, if he was going to take up acrobatics he should wait until his blisters were better, and had he brought his torch with him, he would have seen as clearly as I had that he was wandering off the road towards the ditch. He suggested I had a strange sense of humour and – a few other things that are probably best left to the imagination.

I listened to Pirate Radio, a local station, for an hour or so before sleeping. As was generally the case with the locals we met, the DJ was getting wound up over the eclipse, lack of tourists and forecast poor weather. There were various themed competitions – win a holiday by explaining why you wanted to miss the eclipse – and adverts: 'Blackthorn Cider: sponsors of the eclipse' was my favourite. It all seemed unreal.

I still felt great, my feet were perfect, the split in my left boot was not getting any worse, and I was sure I had lost several pounds in weight.

Phil's blisters aside, his basic fitness was, as he kept telling me, excellent. We were still following the unwritten 'never stop on an up' rule, and so far neither of us had, which given some of the vicious ascents we had made was very impressive.

We were going to have to adjust our first week's target downwards, but with luck Phil's blisters would heal during the following slack period, and we could really go for it in the final week. I would just have to keep repeating to myself: 'The main thing is to enjoy it; it doesn't matter if we don't achieve our targets'.

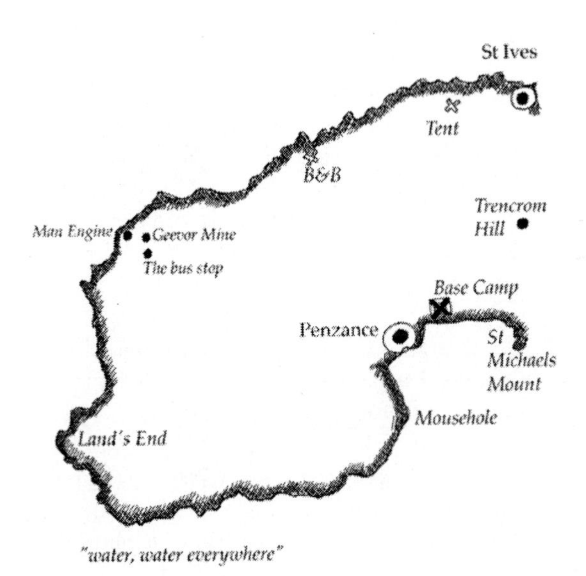

"water, water everywhere"

Pit Stops

Before breakfast, Phil persuaded Mrs Farmhouse to lend us a set of modern scales so we could weigh our packs and I could see how much weight I had lost. On I jumped to discover a disaster: I had *put on* four pounds, and now weighed fourteen stone ten pounds. Phil was heartbroken by this news, nearly choking himself with laughter. I rationalised by persuading myself I had put on eight pounds of muscle and lost four pounds of fat. Now the muscle had been built up, the fat would continue to burn off, so my weight would start to drop. It is frightening how I can convince myself of the validity of such twaddle, then start to turn it into a 'fact', telling other people about it as if it had been scientifically proven.

Next, we weighed our packs. Mine weighed thirty-eight pounds, and Phil's a measly thirty-one, which I took to be fair evidence that the tent should be split between us, or Phil should carry some of my gear. Phil, however, justified the difference in the weight we were carrying in his usual way: firstly, it was roughly in line with our relative body weights, so our muscles were carrying the same proportional amount above what they were used to. Secondly, it was helping me in my quest to lose weight. He also exaggerated his hobble, and I picked up the hidden message: any more weight would obviously put even more pressure on his blisters.

As previously explained, Phil is a keen photographer. I was exploiting this by expecting him to visually record our trip, including the fantastic scenery we were walking through. We had been treated to an incredible variety of top quality sandy beaches, rocky coves, estuaries, hamlets, villages, towns, old mine workings, cliffs (hundreds of these – not surprising for a coastal walk – and had to climb up and down most of them), headlands, farms, woods, bracken, heather, grasslands, cultivated fields, lakes, rivers, streams, gullies, gorges, harbours, granite outcrops, buildings, gardens and ditches. Phil had snapped them all (apart from on Perran Beach where he lost his camera batteries) along with any other interesting sights, such as other walkers, campers, tourists, holidaymakers, Mr (rare) and Mrs Bed and Breakfast, vehicles, furniture… 'You name it, I frame it.'

Now he was in animal mode. No, he wasn't biting people or chasing sheep, although he does like to be called 'the cat', due to his feline qualities, but snapping, with his camera, the farm dog (a beautiful white chow) and cats (a meek and mild black one, and a more feisty, half wild, ginger).

Having eaten a great traditional fried English breakfast, with more eggs if you wanted them – we did – and more toast than you could poke a stick at – although Phil would give it a good try – we said goodbye to the very friendly Mrs Farmhouse and set off.

Why did I say Phil would give it a good try regarding poking a stick at things? He is incredibly focused on exactly what he is doing at any one moment, so much so that he tends to forget e.g. that he has muddy boots on which might be best taken off by the front door, not in the bedroom. Of particular relevance here, however, is his ability to forget that he has a walking stick in his hand and large pack on his back, despite having lugged it up and down hills all day, complaining how heavy it is and how sore his shoulders are 'but not as bad as me blisters of course. You don't know…'

Consequently he is particularly entertaining in those tiny souvenir shops with shelves crammed full of highly breakable, delicate porcelain models of elephants, butterflies, or mock Tudor houses, or pushing his way to the bar to get in our beers. Most of the time I try to keep an eye on him and remind him to leave his pack outside ('Sure, and let someone nick me tent') while dodging a clout around the ear from his rucksack, or a whack or poke anywhere between the shin and eye from his walking stick. The latter is particularly likely if he is taking his rucksack off, because he invariably keeps hold of his stick and it flashes around all over the place while he wriggles out of his rucksack straps and drops it to the floor.

Sometimes I can't be bothered and just let him get on with it, while other times I deliberately hang back and revel in the chaos trailing behind him as people rub various parts of their anatomy or crash into each other in their effort to dodge a poke or clout. No one ever seems to get really angry with him, and he never has the disaster, which seems to permanently loom over him.

We rejoined the official path, begrudging every step down the hill we had slogged up the night before. Incredibly, Phil felt pretty good, and we moved at something approaching normal speed. We could see the lighthouse at Pendeen Watch three miles across the bay, but six for us, and promised ourselves a lovely pot of tea in the café that was sure to be nearby. After several ups and downs to negotiate some coves, we made great time across a two-mile flat stretch. On reaching the lighthouse, we found it to be locked, with no café, despite numerous signs proclaiming 'Cornish teas only two fields away', which we thought somewhat frustrating:

That don't impress me much
So you've got the light
But have you got tea cups
Don't get me wrong
I think you're all right
But that won't keep me warm
In the middle of the night
OOooOOooh Wah, wah, wah

We decided to push on to Geevor Mine, where there is a very interesting and entertaining museum of the Cornish tin mining industry from pre-history until the 1980s, when the mine was closed due to low international tin prices. At first much of it (i.e. several miles of tunnel at various levels down to a thousand feet below the surface) could have been kept open if twenty thousand pounds had been spent on seals and pumps, but the decision to do so was delayed until it was too late and water had taken over. Now they are hoping for a one point two million pound lottery fund grant to reclaim and open a much smaller area.

The 'above ground' site of the mine was in good condition, and it was very impressive looking at the rock crushers and shaking tables. Rock crushers come in two main types, either very large scale mixers in which the rock is forced between metal plates and split apart, or massive tumbling machines in which the rocks bash against each other, the walls, and large iron balls, so as to break them up. (My BR had become only a sore memory, but this sight and the image it conveyed soon brought it back to mind). Shaking tables are very large, flat tables that shake. Having been crushed, the rocks are passed along these tables, which have lines and slots designed to grade the rocks into various sizes.

Other machines included magnets for extracting any metal, and water centrifuges to sort the rocks according to density, which helped identify those containing tin. The scale was vast, with a ridiculously large amount of rock having to be blasted, extracted, crushed, sorted and disposed of in order to pick out the lode bearing material. Disposal tips were everywhere, like man-made lava flows running down the hillsides, but no longer in use and rapidly being reclaimed by nature.

With all this machinery and movement of rock, the noise must have been terrible, but now it was eerily quiet and depressing to recall the effects of closure on the local community. Twenty thousand jobs had been lost at this mine alone, and all around this part of Cornwall, similar mines had existed, closed, and their ruins left to slowly decay. The towns and villages that had supplied the manpower were struggling to maintain a living and their dignity, but everyone we met was friendly and positive.

I'm afraid the beauty and sadness moved me to poetry.

Geevor Mine

When first I set my eyes on Geevor Mine
I felt its peaceful power: a silent shrine
To men who gave their lives to mining tin
Drilling, blasting, crushing, shaking. Din

Iron balls were used for crushing rocks
Hewn by men well used to taking knocks
But now tin's price on global markets falls
Out comes the man from London with no balls

'We have to close the mine' turns to depart
Iron headed man, with iron heart

Thankfully, there was an excellent, though incongruous, canteen: bright, clean, and modern plastic. While Phil tucked into roast beef, veg and spuds, and I had a Cornish pasty, he told me the sad tale of how he came to be living alone.

One day he came home from work and was given a cup of tea and told to sit down by his mum. She then told him that his dad and she had decided they would be selling the house and moving.

'Great, when are we going?'

'We are moving at the end of next week. Unfortunately, there won't be room for you.'

As he told me this, Phil's eyes filled with tears, and his voice began to choke. I sympathised.

'That's awful. How old were you?'

'Forty-two.'

We discussed where to head for that night, deciding on a campsite four miles away, but soon after we set off I had a brainwave.

'Why don't we get a taxi back to Penzance and camp there tonight? We can set up our eclipse base early, and our chances of getting a pitch are probably better today than tomorrow when everyone will be arriving for eclipse week. I will also know where I am and that I can definitely catch my train.'

Phil immediately accepted the logic and magic of this suggestion, so we headed for the nearest village, Hillside, and stopped at a garage to enquire about a taxi:

'Sorry, no taxis around here, but there is a bus to Penzance in twenty minutes.'

Looking round to pass on this good fortune, I spied The North Inn – excellent planning by the lucky fox again.

Having joined Phil in a pint, just to be sociable, I followed him outside to wait for the bus. Since I had the kitty, I boarded first, asked for two singles to Penzance and went to sit down. Suddenly I heard a commotion behind me, accompanied by 'Ow! Watch what you're doing with that bloody thing!' I turned around to see a fellow passenger doubled up with laughter, then noticed the driver rubbing his arm and glaring at Phil. Somehow, he had managed to stab the driver with his walking stick, by exploiting a small design fault in the toughened plastic frame intended to protect the driver from the great unwashed public. Despite this assault, the driver told us where the Penzance campsite was and dropped us between official stops as close to the site as the route allowed.

We walked along the two-lane road leading from the train station, heading out of town towards Marazion. The inland side of the road was lined with old town houses converted into hotels, and the seaward side had a two-foot high wall to deter pedestrians from falling over the edge and onto the rail tracks up to thirty feet below. Presumably the wall was kept at only two-feet high so as not to interrupt the view from the hotels: a few broken heads once in a while seemed a reasonable sacrifice. I was now treated to my first view of Mount's Bay, some three miles of broad flat beach with St. Michael's Mount a couple of hundred yards offshore at the far end. The Mount could be seen to consist of a steep and rocky island, crowned with a high walled castle, with a small harbour and causeway to the mainland at low tide.

We soon found ourselves approaching a roundabout with a large field just before it on the left hand side. In the field there were a few tents and very makeshift facilities, while on the typical farm gate an amateur looking campsite sign invited us to see the site manager before putting up our tent. Phil made a snap decision: 'No way, let's find the official site.'

At the roundabout, official brown tourist signs, with the international camping sign, pointed left, up a side road, so off we went. As we passed a large twenty-four hour Tesco store a thunderstorm started, so we took shelter in the fruit and veg section until we spied the cafeteria, whereupon we bought a coffee and cake and feebly sat waiting for a break in the weather.

Continuing to follow brown tourist signs, we turned left down a narrow lane and arrived at an official looking sign above the gate to... the back of the field we had earlier rejected. Dumb or what? Still, we had avoided the storm and now knew that twenty-four hour food was just round the corner. Searching for the site manager, we approached the facilities to discover that they were even more makeshift than we had first thought, since they were still being assembled, but would be fine once completed. Three concrete steps led up to the floor of the first block, and as I could hear hammering, I decided to stick my head round the corner and let out a 'cooooeee.' This rapidly became a 'coooeeeyowwwaaagh!' as the site manager's Alsatian took advantage of his heightened position and launched a surprise cheek by jowl attack from the doorway. The manager appeared with the 'It's all right, he won't bite you' phrase all owners of savage dogs have to learn by heart before they are allowed to unleash them on the unsuspecting public.

We persuaded them (manager and Alsatian) to let us stay for seven nights, which the manager was initially reluctant to do, claiming to be fully booked throughout the eclipse period, but soon succumbed to Phil's look of desperation, extreme limp, and a slight oozing of the blister story. We soon had him taking pity on us poor fools, giving us a very nice space along the back edge of the site, within a two minute hobble, say, twenty-five yards, of the shower block, close to the back gate for convenient access to our food source, and nicely shaded by trees. For the whole of our time there, he really looked after us, checking on the progress of Phil's blisters and asking me to go easy on him / them, advising us where to eat and what to see, and generally making us feel welcome.

Having erected the tent, we relaxed for a couple of hours, although I was already having second thoughts about how good the site was, due to the lumps and bumps sticking in my back, before going to a recommended local pub, Coldstreamer Inn at Gulval, for our automatic fish and Guinness. To get there, we had to 'go out the back gate, left at Tesco's, through the council estate and down a footpath across some fields heading for a church, then up the hill, the pub is on your left.' Phil had renewed his plasters, and may even have washed his feet, but even so was a sorry sight as he lent wearily over his walking stick and limped along wincing at every step, like Gandalf after his confrontation with Saruman. To make him feel better, I ran up the hill in front of the church.

Phil took one look at the menu board and started to salivate.

'Sea bass. I can't believe it. Sea bass. That's the best fish there is. No point in reading the rest of the menu.' As I'm writing this, I am beginning to realise how much I allowed myself to be influenced by what Phil said and did, and this was another example.

Being aware of the value of money (tight as the skin on a sausage) I had to force myself to commit eleven pounds and fifty pence to a fish, even with Phil insisting it was one of the best fish there is and that it is not often available so should be eaten at every opportunity. When it came to paying the bill, however, I was very pleased to find it only totalled twenty-three pounds between us, despite both also having had a three pound fifty starter, three beers and a coffee. Due to some failing in my up bringing, my immediate elation gave way to pangs of guilt: I just couldn't pay the bill and scoot, so asked whether we had been correctly charged. 'Don't worry, sir, you've got a bargain,' said the extremely nice manager. Result! I had the bargain and no guilt pangs.

By the time we had retraced our footsteps to camp, it was half-past eleven. I don't know why, something to do with minor twinges in the area below his ankle, I think, but Phil decided he didn't want to walk into downtown Penzance to see what was happening, and would get some sleep instead. I decided to go for it.

As a teenager, my dad was in the sea cadets, and during the war, in the merchant navy. He misses few opportunities to tell me tales of sea life and daring do, and I have picked up a strong affinity with the sea, so I spent a thoroughly enjoyable half hour or so in the harbour area, inspecting the ships, watching the yachts bobbing and waves lapping at the outer wall.

Being alone and far from home, I guess it was natural that, while walking, I recalled one of my favourite tales, about how dad's life had been saved by a mosquito. As a boy, he had suffered badly from osteomyelitis, which had necessitated him having an operation on his left leg. Most of the muscle and flesh on his inner thigh had been cut away, and the wound had been packed with a dressing, which was to be gradually reduced as the flesh and muscle grew back. The wound had been extremely itchy, however, and my father had laid hands on a knitting needle and poked it down the bandages to scratch the leg. As a result, when the bandages were removed, the packing had all been pulled away from the wound and the skin had grown over, leaving him with a permanent oval hole in his leg some eight inches long and four wide, as though someone had sculpted a rugby ball out of it. His tibia was clearly visible beneath the skin, running along the length of the hole, half an inch above the surrounding flesh. Immediately after the operation, his left leg was two inches shorter than his right one due to the loss of muscle and tissue, but he managed to obtain a bike and spent all his time pounding away on it until he could walk straight without having one foot in the gutter.

During the war, he sailed out to Canada as a member of one of a number of crews going to pick up new ships. While working on his ship to make it seaworthy he was bitten by the aforementioned mosquito, and had an osteomyelitis-related reaction, which left him in hospital when his ship sailed. When he eventually made his way back to England and went to collect his pay, the pay clerk was shocked to see him and informed that his ship, *WC Teagle*, had been torpedoed. Since it was an oil tanker, it had sunk with all hands: my dad and the Chief Cook, who had also been in hospital, were the only 'survivors'.

When he arrived home and knocked on the door, his mum almost fainted as she had assumed he was dead, having been given the traditional notification: firstly a letter saying, with no stated reason, that she would no longer be receiving his pay, then a week later another stating he had been reported 'missing in action'. I guess there are many ways to break bad news, but this seems rather similar to the headmaster who is supposed to have adopted the method: 'All those with mothers, one pace forward. Where are you going, Jones?'

I must have been told this tale when very young, because my version was that my father had been saved from a burning ship by a mosquito flying him to safety.

Following a very quick look at the dockside part of town, I walked back to camp along the same road for the third time, although it was much more interesting now as it was filled with groups of well-laced people obviously returning from a night out somewhere the other side of our field. They were all in extremely good spirits, presumably due to the extremely good spirits in them.

On the site, all was quiet with very few people still up – even the ubiquitous 'sing along to John Denver's greatest hits' guitar player had turned in – and the interesting smells and smoke trails that had been wafting from virtually every tent, had gone.

In the morning, I lay in my sleeping bag listening to what sounded like hundreds of new arrivals busily erecting their tents, so was surprised to find the site still only a quarter full when I went for my shower. Erection of the facilities had been completed the previous evening, under the watchful eye of the owner's Alsatian, and there were now hot showers, hand basins and mirrors. Quality is best illustrated by the fact that the mirror was made of polished tin.

I was due to catch the 11.40a.m. train to London for Sally's twenty-first party, so following a roll and coffee from Tesco's, packed the things I wanted to take home, including all my clothes, for washing, swapping, or at least fumigating. I decided to leave my sleeping bag, ground mat and most other gear in the tent, hoping that 'after two minutes some ****** would *not* nick it'. I would only be gone two days, and Phil would be there most of the time, though I realised that was a mixed blessing.

I had invented a fantastic scam with the take-away coffee, so if I tell you, you must promise to keep it quiet. The drinks machines were regulated to fill a standard cup, and the space was too small to fit a large one under, so I just kept refilling a standard cup and pouring it into two large ones until they were full, with three standard sized coffees in each. The serving wench only charged for a standard take away coffee, so I was quids in. Of course, if I had called her a serving wench at the time, she would no doubt have charged me for six, and I would have ended up wearing the coffee. Why didn't I feel guilty about robbing Tesco like this, having confessed when undercharged in the pub the previous night? One of life's mysteries.

Phil decided to accompany me to the station despite still being in great pain. I was worried that this would look too much like daddy seeing Little Johnny off on his holidays or some other, even worse interpretation, but couldn't help being impressed by his genuine concern for other people's well being. There would have been no chance of me seeing him off if the blister had been on the other foot: I would have expected him to fetch me a Tesco weekend hamper special, then make his way to the station on his own, taking my dirty washing with him.

While Phil went through the fifteen-minute preparatory ritual prior to donning his boots, I suddenly realised, with my usual level of observation, that despite the pain they caused, he seemed to wear his boots all the time.

'Haven't you brought anything more comfortable with you?'

'No, you said keep the weight down, so I only brought one pair of boots.'

'Nothing for wearing round camp, or in people's houses?'

'You said keep the weight down.'

In line with my natural generosity, I asked if he wanted to try my walking sandals, which I had hardly used since the steam train trip (could that really have only been five days ago) when they had given me what I had at the time called blisters but now, having seen the real thing, regarded as minor irritations. The split near the toe in my boots had worsened slightly, and I had decided to prepare for their disintegration by taking the sandals home to swap for my old worn-in boots. If Phil borrowed the sandals, it would save me lugging them back.

As soon as he put them on, Phil started to rejoice. 'That's fantastic. These are miles softer than my boots. Of course, I am still in pain, but I can put up with this.' He looked weird in sandals with socks (one of my pet hates) but weird and Phil go hand in hand, and if it helped him cope, I wasn't going to complain. He still had a marked limp / shuffle and attracted plenty of attention from people who probably thought he had some incurable crippling disease, rather than that he was on a walking holiday for pleasure.

There was a young couple in the tent next to us, and just as we were getting our last few things together they finished putting the chain back on one of the two mountain bikes they had hired for the weekend, gave us a wave and set off. We had decided to pop into Tesco's since Phil wanted to buy some film, so soon followed our neighbours through the back gate and onto the path, where we found them trying to mend the brakes on one of the bikes. They were the first people we had overtaken since the guy with the bad knee at The Carracks.

As we were about to enter Tesco's, a shopper bus stopped next to us. Phil's eyes lit up:

'Where you going?'

'Town.'

'When?'

'Now.'

'How much?'

'Free.'

We have had such amazing luck. If Phil hadn't wanted the film, we would not have known this bus existed, and would have spent ages struggling over the boring hill to the station. In practice, it was ten minutes before the bus left, full with piles of shopping bags, and the occasional arm or face sticking out. As we pulled off the roundabout and headed into town we saw our neighbours on the grass beside the road, this time pumping up tyres.

In town, Phil bought a pair of black walking clogs and returned my sandals to me for de-lousing. The clogs were not the Dutch wooden type, though that would have been reasonable given the number of Dutch people we had met, but walking shoes with hardly any back and no laces, which made them look incredibly loose but he swore they were fine. I bought a Cornish pasty to eat on the journey, and we had another breakfast, during which Phil apologised for his performance so far!

'Listen, John, I know I have let you down by getting these blisters. You set the target of Penzance by the end of the first week, and I have blown it for you.'

'Don't be daft.'

'I'm serious. We came on this walk together, having agreed realistic targets, and I've let you down. I should have made sure my feet were tougher.'

'Come off it. There was no way you could have known you would get blisters like those. They were incredibly bad, and are still rather gruesome now. I've enjoyed the slower pace – it's allowed me to see the sights better.'

'I should have been more careful, insisted on stopping sooner on day one.'

'You were just unlucky. It could have been me with my BR.'

'Look John, let me tell you something. Remember Kilimanjaro? How gutted you were when you failed the first time, so much so that you turned round and went straight back up?' I nodded. 'Well, you weren't the only one who failed. Remember how I gave up on day three?'

'Phil, you had the worst case of Kili Killer Kwik Klik Klip Klop Klap Trap Splat Krap they had seen since the war – bad enough to make a camel dump its hump. There was no way you could have possibly carried on. In my book, that doesn't count as failure. Failure is only when you don't try your hardest. I could have made it to the top first time, but hadn't properly prepared myself for the mental challenge, so that was genuine failure. There was nothing you could have done to prevent what happened to you, or fight your way to the top once it grabbed you.

'There weren't any bags of quick set cement lying around, or a giant corking machine, and even if there had been, sooner or later something would have blown, with disastrous consequences for you and anyone else within a half mile radius. Anyway, it was the way it just took all your energy that stopped you even before the pebble dashing started. I remember how you seemed to fade before our eyes. One minute you were setting the pace, the next you were hobbling along like someone on the South West Coastal Path ridden with giant blisters, and a few minutes later you were having to lean on me for support even on that long downhill bit to the lunch stop.'

'At last, you've hit it! I agree there was no way I could have carried on up – I only just made it back to Horombo Huts as it was, and still took two years to fully recover. In my head I agree with you that it wasn't a real failure, but in my heart I still want to return one day to prove to myself I can do it. If you do ever write this walk up and call it Mister Blister, I hope your next book will be 'Master Blaster, Phili Kill Kili'. But that isn't the failure I'm referring to.'

'So what are you on about then?'

'The last part of what you said. It was you who tried to help me carry on. There were six of us on that climb, and I had only known you for six months, whereas I'd known the other four for donkeys ages. We'd worked together for years, and been on all sorts of walks together all over the world, but not one of those bastards made the slightest effort to help.'

I knew he was right, because I had been surprised, even shocked, by it at the time. They had all basically ignored his suffering to concentrate entirely on their own personal success. It had been the first time I had seen any of them under any real pressure, and it was clear to see they were effectively just a bunch of individuals walking together for company, and definitely not a cohesive group determined to see success for all members. This had been confirmed very dramatically on the final night climb to the top, when each of us had just done our own thing – yes, I was as guilty as they were – only one of us making it to the top, when at least three, and probably all except Phil, could have done so. It had been in stark contrast to the way the far less physically fit Japanese all made it by helping each other.

Despite enjoying the heroic image of myself, I felt I had to defend them against Phil's criticism, and try to lighten the mood.

'Don't forget it was tough for all of us. Besides, if you hadn't given up at the lunch stop I was going to push you over the edge to ensure you didn't drag me down with you.'

I could have added that, having known him much longer than I, they had probably had experiences similar to those I was going through every time he whined on about his blisters. On several occasions, I had resisted the temptation to shove him off a cliff, so they may have had similar mixed feelings when he started to complain incessantly how tired he was, it was okay for us, we didn't have to listen to people snoring all night. I continued to think on these lines while Phil, ignoring my frivolous comment, explained his feelings further.

'The failure I am talking about is their failure to help me. Ever since then, I've felt differently toward them. Sure, I still go walking and drinking with them, but Kili was a real eye-opener. I wouldn't trust them to help me in a tight spot – instead of giving a helping hand, they just turn a blind eye. You were the only one who showed any concern for me or made any effort to help, so when we arranged this walk, I swore to myself I wouldn't let you down.'

By now, I was hardly listening to Phil, attempting instead to recall what it must have been like for the other four as Phil worked himself into victim mode – '...you don't know the pain I'm in... lack of sleep... well known torture... stomach pains... it's agony... it's not the fitness... so tired... oh, the cramps... it's alright for you... I'm as fit as anything... listen people... headache soon... you don't get pain... I can easily cope with the up... lack of sleep head... I'm as fit as anything... it's these snorers... you don't get the pain I'm in stomach aching every step it's fitness agony... oh the head... double up fits of cramped sleep... all for the right you're snore... easy up in right agony... stomach the pain the snorers ache it's a head pain... I'm fitting no pain stomach... you fit no stomach head pain... know up feign pain again... brain ache make awake... take pain... last straw snore... I'm for you in right any pain thing... I'm a pain in the arse...' whine, whine, moan moan.

My recent experiences brought this awful image to life, and I could easily see why, given the difficulty of their own struggle, they had kept well away from Phil's dying swan routine. Any heroics on my part had largely been through ignorance, which didn't fit with Phil's interpretation.

'I hadn't realised how strongly you felt. It really isn't that important, and I'm sure you have got it wrong. They were probably so focused on their own efforts that they didn't realise how bad you were, or when they saw I was helping, decided they didn't need to.'

I didn't add 'and given the same situation now, I would be the first to wave you goodbye.' Luckily, Phil couldn't read my mind.

'It was thinking how you had tried to help me despite your own tiredness which kept me going through the blister days, and I really just wanted to say that my blisters are getting better and I am determined not to let anything else get in our way next week.'

Suddenly I was late for the train, so had to make a dash for it, with Phil's words ringing in my ears:

'I won't let you down. I am determined not to let anything get in our way next week.'

As I fought my way through the crowds of people arriving for the festivities, I kept wanting to apologise, justify my departure, and explain that I would be coming back in a couple of days. Even for me, taking a holiday from a holiday to attend a party was a bit special, and it felt odd to be sitting on a virtually empty train taking me away from all the action building around the eclipse. Nevertheless, I enjoyed the relaxation of the journey to Paddington, and even the hustle and bustle across London.

With no time to go home first, I had arranged with Lynn that she would bring some spare clothes for me to change into at Sally's house, so my arrival caused much merriment and comments along the lines of 'it wasn't supposed to be fancy dress, but it was a great idea to come as Robinson Crusoe.' The party was brilliant, I only mentioned Sally's Bottom a few times, and it was great to see so many friends and relatives. Most did not believe I had returned from Penzance specifically to attend, and was really going back again two days later.

On Sunday, I relaxed, changed my gear, repacked and told my family tales from the first week. It was great to be home, but I felt in limbo, so when Monday morning came I had mixed feelings – it was both 'too soon' and 'at last'. The strangeness was heightened by the feeling of déjà vu as I walked to the station, although being 7.30 on a Monday morning, the roads were now full and there were crowds of workers rushing for their trains to start another week's hard labour – just another manic Monday. Some gave me quizzical, mostly jealous, looks, and I certainly didn't want to change places. I hoped to see my fox, but was disappointed, then persuaded myself that this was because he was already waiting for me in Penzance.

Having booked my seat from Paddington in advance, I was able to sit back and enjoy the entertainment of crowds pushing on at every station, then having to squat in the corridors or between the seats, or just give up and stand. My only fear was that my seat would be snapped up as soon as I went to get a coffee or pay a visit, so I struck a deal with those in nearby seats and we guarded each other's.

I arrived back in Penzance, with a kit of clean clothes and two pairs of well worn-in boots, at three-thirty in the afternoon. A couple of day's rest had left me relaxed and refreshed, but dead keen to start walking again and get away from the masses. Penzance was now quite crowded, it being only two days before the eclipse, and there was a general buzz of anticipation in the air. Many locals were, however, already criticising the way the event had been handled, and I was actually surprised that there were not more people around. Throughout the summer, rumours had abounded as to how many people would be descending on Cornwall, with the person appointed by the county council to manage eclipse marketing adopting the interesting tactic of predicting up to three million visitors and complete chaos on the roads. Numerous major events had been planned, with all the best viewing points snapped up by people hoping to make a killing charging exorbitant prices to thousands of hapless visitors, or by New Age travellers as we had seen at Newdowns Head.

During the first week of our walk, locals had been saying that tourist numbers were, if anything, low for that time of year, and had been all season. Various reasons were put forward, including regular visitors being put off by the predicted crowds and deciding to go elsewhere for a change. Others might have decided to move their normal week so as to coincide their holiday with the eclipse, but the effect of this could not be judged until the event. Many people, whether regulars or eclipsologists, would adopt a 'wait and see' approach, deciding what to do at the last minute based on the weather forecast and actual crowd levels. Everyone agreed that good weather would be vital for attracting anything like the predicted numbers, and turning what would otherwise be a poor season into the predicted bonanza.

For some time, the weather forecast had been getting worse and it was now predicted that there was only a twenty percent chance of clear skies over Cornwall on eclipse day. AA reports said that very few people had travelled down at the weekend, and it was recognised that, failing a last minute rush, which would cause horrendous traffic problems in itself, the numbers were just not going to be there.

I pushed past the opportunists trying to persuade people arriving by train to part with sizeable wads of cash in return for doubtful promises of accommodation of dubious quality, which would not be available through any other outlet at their 'give-away' prices. I knew my accommodation was secure unless 'some ****** had nicked me tent' so headed along the same old bit of road to the campsite, only marginally concerned as to what I would find.

Finally, the owner's prediction had come true, and the site was jammed full. I was soon able to make out our tent, however, and it looked fine: at least it was there and standing upright. On entering, I was immediately struck by an awful smell, so carried out a general investigation to discover its source, fairly quickly locating a carton of putrid milk and some other suspicious items Phil had presumably used during the previous morning's breakfast and neglected to clear away. Otherwise, the tent had been left in generally good nick, so in all I was pleasantly surprised, and relieved that I had a roof, albeit a canvas one, over my head for the night. After a while, I noticed the smell was still there, and eventually discovered the real cause. I won't bore you with the details, but it is easy to see why Phil lives alone.

I tried to force myself to laze around for a while, but became increasingly restless, having spent much of the day just sitting on trains. Soon, I found myself heading back into Penzance, but there didn't seem to be much happening, and I allowed myself to be tempted by the advert outside the cinema to watch *Austin Powers – The Spy Who Shagged Me*. The show started at 5.45p.m., but for some reason which I could not fathom, they would not sell tickets until fifteen minutes beforehand unless the purchaser agreed to stay on the premises for the whole time between buying the ticket and the show commencing.

It was now five, so I agreed to stay, bought my ticket and left to do some window shopping, returning five minutes before kick off to find chaos all around, but no problem as far as I was concerned. I was disappointed with the film, finding it slow, with too many feeble, telegraphed jokes, and making unnecessary and excessively offensive digs at fat people. When I told various friends and acquaintances of my reactions, they said 'what did you expect? It was just like the original' so no sympathy there.

By 7.30p.m., I was on the street again. I fancied something to eat, but thought Phil might be back from his Scilly Isles trip by now, so decided to nip back to camp to see if he wanted to join me. I had just climbed over the front gate when I spied Phil shuffling slowly towards our tent. He was clearly still suffering from his blisters, but assured me he had been bathing his feet in the sea at every opportunity (the owner had said, as most people would guess, that the salt would be good for them). They were feeling much better and he was confident of being up to speed by the time he returned from his trip home to watch Wolves play Wycombe.

It took Phil about two seconds to notice my haircut. 'Wow! I can't believe it! That is short! Whatever came over you?' which was fair comment: I had gone for a 'number three', carried out with much hilarity and pleasure on Sunday by my son, Peter. Had it been Phil with the new haircut, and me commenting, I would have followed up the 'Whatever came over you?' line with something like 'A combined harvester?', but Phil is just not like that. He takes all the stick I, and anyone else, give him, and never retaliates, being one of those good-natured people that sarcastic beggars like me thrive on.

We returned to Coldstreamer Inn, where I had Moules Marinière for my starter again. On both occasions they were fantastic, a vast pile of mussels in a rich sauce, with a large chunk of brown bread and butter. Phil switched his starter to Chicken Bang Bang and had Mediterranean Monk Fish main course. I broke ranks by having a wonderful T-bone steak, and we really pushed our rebelliousness by having a carafe and a half of wine instead of our usual Guinness (well, we did have one pint each). Unfortunately the bill was correctly totalled this time, but we were full and more than satisfied.

During the meal Phil told me that the Scilly Isles were 'a top place' and I should take some bird there: 'Island Hotel on Tresco is a fantastic place to take a wild woman. Couldn't fail.' Since the only bird I would want to take there is my wife he could be right: I should at least stand a good chance. 'Fantastic beaches with deep blue sea and semi-tropical plants.' On the crossing he had helped out an American woman who was seasick. 'The boat was going all over the place, up and down, side to side. Everyone was ill, except me – it doesn't have any effect on me, I was walking up and down the deck loving it. Anyway, I got her a sick bag and sorted her out, then she latched onto me for the rest of the day. She is on a three-week trip from New York, and should have been with a friend, but the friend decided not to come at the last minute. She nearly cancelled as well, but decided to come alone. I told her not to be angry with her friend, and that she would be better off without because she could do exactly what she wanted and would meet more people. If she hadn't been on her own, she probably would never have met me. She thought it was a top place, and if she ever meets a man, she is going to take him there. Can't fail.'

I resisted the obvious comments (how lucky she *wasn't* to have met him, and that it was hardly a compliment when she said to him '*if she ever meets a man*') and told him about my twenty-first birthday party instead.

The Sundance Kid

Only one more day to the eclipse. Although it would be total for two minutes where we were, predictions of the likely atmospheric conditions were now even worse than the ones of the day before: we had only a ten percent chance of seeing it because a blanket of cloud was expected to cover Cornwall at that time (eleven minutes past eleven on the 11th of August). If we had stayed in London, we would have had a fifty percent chance of actually seeing a ninety-five percent eclipse.

Was a ten percent chance of seeing a hundred percent eclipse better that a fifty percent chance of seeing a ninety-five percent one? Is a ninety-five percent eclipse ninety-five percent as good as a full one?

We had both bought a pair of eclipse shades advertised with the impossible to prove or disprove claim 'the best selling solar viewers on the planet – safe solar viewers for the 1999 total eclipse'. Whether we used them remained to be seen, but we didn't want to risk being caught without a quality pair of shades if the clouds did part. Warning slogans played on the radio such as 'I'm going to see this eclipse if it's the last thing I do' were strong reminders of the danger of using anything but a proper solar viewer. I could not see anything at all through them, except when looking directly at the sun, when I could see a rather faint image. Even so, I was still worried that ultra violet rays would get through and leave me permanently blind. The logical thing was to watch the eclipse on television, but I guess most people don't act logically in situations like that, and I knew I certainly wouldn't.

We struck up a conversation with three lads from Blackburn, brothers Arno and Greavesie, and Tristan, the latter, younger brother's mate. They had arrived late the previous evening and were now lying half out of their sleeping bags, which were half out of their tents. They had already opened a can of beer each, and we were to find that they spent the whole time drinking, smoking, coughing, laughing, or some combination of the four, being fantastic company and game for anything game for them. They had all the gear, having driven down in Arno's Range Rover, and as soon as Greavsie struggled out of his sleeping bag to put the kettle on, Phil successfully worked his 'I could murder a coffee' trick, more to keep his hand in than out of any real desire since we had each just finished one of my 'three for the price of one' fiddle take away coffees.

Arno, being the eldest and owner of all the gear and motor, was the most responsible one, tending to mother and generally organise the others. We initially assumed the brother's names were based respectively on a marked similarity in physical prowess and stature to Arnold Schwartzennegger, and football skills of Jimmy Greaves. This illusion was shattered when Arno crawled out of his sleeping bag to reveal a well developed beer belly and height of around 5'8", and Greavsie attempted some flash move with his empty beer can, only succeeding in kicking it hard against the door of the Range Rover and bringing joyous comments from Arno, including reference to the distance back to Blackburn on foot. The name Tristan conjured an heroic, Greek God image, and he lived up to it as much as his companions did theirs, being extremely tall, skinny and sickly pale.

While drinking the cadged coffee, Phil fretted over what to do. There was never any question about his not going back home to watch Wolves play Wycombe, but he could not decide whether to drive back to Penzance overnight, or catch the only available train back, midday from Paddington. Driving back would mean he should get back in time to see the eclipse, and we would have his car available for the return to London at the end of the walk, but he would have to find somewhere to park for the duration, and would be very tired after driving through the night with, possibly, hundreds of thousands of other last minute 'late arrivals at the eclipse ball'. Taking the train would mean travelling throughout the eclipse period and only seeing a ninety-five percent eclipse at best, but he would arrive back fresh and rested.

Phil went back over the argument yet one more time. If he could be sure that driving back would mean seeing the full eclipse, he would do so, but with the forecast likelihood of seeing it being only ten percent, was it worth the effort? He had a much better likelihood of seeing at least ninety-five percent in London. Eventually we decided to keep it flexible: he would arrive back whenever, using whatever travel mode, and we would meet at the tent sometime. That is the kind of arrangement I like best.

It was Phil's turn to pack his dirty washing, risk having his other gear nicked, and set off for the big city. What a crazy few days we were having. I started to plan my day while discussing our chances of seeing the eclipse with the Blackburn boys.

On my free day from the walk, I decided... to walk St Michael's Way. That doesn't mean I would be carrying a cross, stopping every 100 yards to pray, or just adopting a variation of the Monty Python silly walk. St Michael's Way, as you may recall, is a walk from St Uny's church in Lelant to St Michael's Mount, Marazion in Mount's Bay. I would be doing it backwards (the walk, not the walking. That wasn't St Michael's way either) starting at St Michael's Mount, and going across the Cornish Peninsular northwards.

Good planning meant that the tide was out so I could enjoy a pleasant stroll along the sand for the two and a half miles from our camp site round Mount's Bay to St Michael's Mount, thereby saving half a mile compared to the high tide route. I stopped to inspect various rock outcrops only accessible at low tide, and wonder how they obtained their names: Long Rock, Great Hogus and Chapel Rock. I walked across the causeway and through the harbour entrance, climbed some slippery steps onto the harbour wall, and followed it to the small collection of shops at the foot of the hill. I felt I had earned a Cornish ice cream, and had a quick mooch around while eating it. I discovered that, being a member of the National Trust, I was entitled to free admission to the Castle at the top of the Mount, which greatly pleased the accountant within me.

Following the very pleasant cobbled path to the top, stopping halfway to inspect the 'giant's heart' cobblestone, I discovered a long queue of people waiting to enter the castle. Highly disappointed at giving up a freebie, I decided not to wait, so contented myself with admiring the views from outside. Looking west I could easily see Penzance and then places heading off towards Land's End: Newlyn and Penlee Point on the east side of Mousehole, with St Clement's Isle off the coast. I knew that, God and blisters willing, we would be coming that way in a few days.

To the east I could see where we hoped to continue: Cudden Point off Prussia Cove; Trewavas Head; past Portleven Sands and on towards the Lizard: what pleasures did they hold in store?

Inland, I tried to make out the route of St Michael's Way, but must confess failure, so decided to crack on with doing it on the twelve inches to a foot scale. I retraced my footsteps down the Mount, back across the causeway, and into the small town of Marazion.

Between St Michael's Mount and Ludgvan, there is a choice of route. One goes via Penzance, starting with the two and half-mile beach walk I had just done, so I decided on the other. Officially this goes to Ludgvan via four miles of road, and even by substituting footpaths wherever possible, I was unable to avoid three miles of the dreaded black stuff. Given the abundance of potential good walking routes all around, this was particularly galling and hard to understand. Surely everyone would be happier with the walkers in fields, leaving the roads for vehicles.

As I *passed* a pub on the outskirts of a small village named Lower Quarter, I noticed there was a group of people sitting outside enjoying a sumptuous feast. A man was wearing a set of (presumably) false horns and a Viking helmet, so I took it to be his birthday or something. I had my 'nutter' T-shirt on, and when one of his companions saw me, she called out 'You should be wearing that.' Cheek. As if I would wear stupid looking false horns and a Viking helmet.

I took advantage of a shortcut path to join the main St Michael's Way at a footbridge just before Angwinack. The hill leading to Boskennal was covered in a crop, which, although clearly not yet fully-grown or ripe, was already at least seven feet high and quite dense, consisting of a long, thin central stalk with short branches slanting sharply up towards the light. No space had been left for the path, and, although it had been fairly well used, it was spooky pushing through the dense growth, particularly when stems seemed to deliberately trip me up or wrap around my legs. As soon as I tried to speed up (only so that I could have a longer rest at the top of the hill, of course) my legs became even more entangled. I can see how a certain kind of person could really panic in such circumstances, and was immediately reminded of the time my uncle Jim went to see *The Day of the Triffids* at the cinema, and at a particularly scary part, placed a stick of celery on the shoulder of a complete stranger sitting in the row in front.

Soon I found myself at a ford, with a number of signs pointing to the 'Sun Block Total Eclipse Festival Site', which was on a large area of flat ground between Cucurrian and Higher Cargease. My route passed through several fields, which had been staked and taped out ready for the anticipated influx of eclipse seekers: they were almost empty. At Trembethow, I had one of those encounters with a herd of cows that tend to put off townies like me: I had to climb over a gate and push through them all as they were gathered waiting to be milked. I am always a bit iffy in that kind of situation: I don't like the way they look at me, and am concerned that they could panic and crush or kick me to death accidentally. I suddenly realised I was turning into an incredible wimp. What chance is there of me ever tackling the Appalachian Trail with its wild bears if I am scared of seven foot high vegetables and a bunch of domesticated cows?

With thoughts like these passing through my head, I arrived at Trencrom Hill. It was obviously a good viewpoint, so I had to climb it, and was delighted with the result. Looking south I could see St Michael's Mount, Penzance and the rest of Mount's Bay, while in the north were such reminders of the previous week's walking as St Ives Bay and the three mile sand beach back to Godrevy Point, which we had skipped by taking the inland road route, thereby making Phil's feet even worse, and probably where he lost his sleeping bag and mat. I could clearly see the dunes we had struggled through the following morning, and way back beyond them, St Agnes Head where we'd had our first night's stop and the life-saving taxi had appeared out of the mist. At the near end of the three-mile beach, I could see Hayle, with its awful six-mile detour but great Philps bakery where we had stopped for tea and cakes, and Phil's guardian angel, Sister Blister, had worked her magic. St Ives was hidden by intervening hills, but I could picture it 'just there', as I could picture Birgide and Ingrid, having accepted that I was less likely to see them again than I was to see the eclipse.

I decided to have my lunch, then lie back on the grass and dream on for a while. Two hours later, I woke with a smile on my lips, peace in my heart, and a slug on my neck.

All my conscious time on Trencrom Hill, there had been a steady flow of people checking it out as an eclipse-viewing site. It did seem to be ideal, with north and south views of two dramatic coastlines, and land views between from west to east ideal for watching the effects of changing light on the colours of the fields, woods and rocks. It was also high enough to see the moon's shadow race across the ground, which I had heard was one of the most fascinating and awe inspiring aspects of a full eclipse. It was only two hours walk from the campsite, so I resolved to return early the next morning to claim my spot.

To complete St Michael's Way would require walking down into Carbis Bay, then along the coast past Carrack Gladden rocks and Porth Kidney Sands to St Uny's church at Lelant. Despite the possibility of meeting my mermaid again, I was in a 'been there, done that' mood, and I decided not to bother – a target given up, just like that. Instead I headed back for Ludgvan, intending to take the Penzance branch of the Way there. I was determined to be brave, but my resolve was not really tested since the cows were nowhere to be seen and the triffids were far less spooky when approached from above, as I could see over their tops.

Congratulating myself on my bravery, I crossed a footbridge and pushed through a hedge to find a woman, clad in a long white robe, with twigs in her hair, standing perfectly still and apparently staring at me, or more precisely, through me. As I approached, I said 'hello' and a couple of the standard phrases used when passing strangers, but she continued to stare straight ahead as though in a trance. Knowing there were some strange people around, and not particularly wanting to be a sacrifice to the Eclipse Goddess, I made a swift exit casting frequent glances back over my shoulder, half expecting to find a band of Druids coming after me brandishing clubs, fire brands and stone knives. Appalachian Trail – come off it.

As I neared Penzance, the sea was just too inviting, being completely calm and with just a few people paddling or sunbathing, most having already packed up and gone back to their hotels, tents or wherever. I stripped off and had a dip in my rather fetching underwear. It was magical, immediately taking away all stiffness and aches from my legs, so I just floated around for a bit, did a few lazy strokes, floated some more, and watched the reflection of the sun on the water. Having no towel, I sunbathed on a rock for a while to dry off, but when I put my shorts back on the wet came through. Anyone who saw me walking along with wet patches in my shorts on Tuesday, the 10th of August 1999 in Penzance, the above is the correct explanation, and any rumours to the contrary are entirely false.

I decided to have a cold salad for my evening meal by my tent. I stocked up with a French stick, mushrooms, tomatoes, corn on the cob, chicken slices and two bottles of wine, reasoning that in the unlikely event that I didn't drink both bottles that evening, I would have one to toast the eclipse with. I was confident that someone would have a corkscrew, and sure enough Arno had a penknife with a corkscrew blade – but nothing for getting stones out of horses hooves, which I thought was compulsory. I spent the evening with the Blackburn Boys, swapping tales, drinking bottles of Becks (them) and wine from a yoghurt pot (me). I made them a chicken salad roll each, and still had plenty for myself, and we slowly wound down into a nice quiet night of semi-drunken stupor, enjoying the relaxed friendly atmosphere surrounding us.

One of the lads noticed several police vehicles and an ambulance, lights flashing, in the road outside the front entrance to the campsite, and in that morbid fascination, which tends to overtake people in such circumstances, we wandered down to have a 'sticky beak'. Some poor lad of about twenty had been hit crossing the road and was still lying there. We watched as the ambulance crew wrapped him up and lovingly placed him on a stretcher and into the ambulance.

Just when you are enjoying yourself and let your guard down, reality always finds a way of creeping back in. Phil's blisters and our concerns over whether we would see the eclipse were suddenly put into their proper context. I hope he was okay. We returned to our tents, finished our drinks and turned in.

Wednesday, 11th August. *Eclipse Day at last!*
In the Penzance area, the moon would start covering the sun around 9.50a.m., full eclipse would last for two minutes from 11.11a.m., and the last 'contact' would be around 12.30p.m.

People had waited years and travelled hundreds of miles. Bring on the show.

I woke at 7.30a.m. to find one hundred percent cloud cover. Not a good sign, but I decided to get up, have a proper look round and see what the latest forecasts were before deciding whether to carry out my intention of returning to Trencrom Hill.

The Blackburn Boys were in customary positions: half-out of their half-out sleeping bags. As soon as I stuck my head out of my tent, they greeted me with 'No luck, John. Been listening to the radio. We are going to have full cloud cover throughout the eclipse, with frequent showers just to rub it in.' We began to exchange views on how unreasonable Mother Nature was being, although I knew that if I was in charge, I wouldn't be able to resist making it cloudy just to show who's boss and have a laugh at the expense of all the people who had put so much time and effort into seeing the eclipse. Mind you, I would probably relent at the last moment and clear the clouds away, so I hoped that was what would happen.

It was soon too late for me to get to Trencrom Hill for the start, and in any case, I didn't fancy two hours walk both ways just to stand in a large crowd of people watching a cloudy sky and getting soaked, so decided to stay at the camp and wallow in self pity instead. A sombre mood could be felt throughout the campsite, and most people with vehicles were jumping into them and setting off in search of a break. Periods of slight lightening of clouds (hurrah) were followed by darker clouds (boo), and then down came the rain at 8.45a.m. There now followed periods of showers and very heavy cloud mixed with some lighter cloud, but always cloud.

The lads invited me to join them, driving around to look for a vantage spot, but I decided to stay put. At 10.30a.m., the campsite manager took pity on us and made his way through the site telling people he would lead all those interested to the top of the hill at the back, which was a very good viewpoint, hidden from the campsite by trees, and only a ten minute walk. Some thirty people, most looking rather unhappy and frustrated, including me, emerged from their tents and followed him out the back gate, along a bridle path, through some fields and up on to what turned out to be an extremely good vantage point. We could see all round Mount's Bay from Mousehole to Cudden Point, with St. Michael's Mount prominent. It had stopped raining, but the cloud cover was still total, although there was a narrow strip of relatively bright sky along the southern, seaward, horizon.

Two of the crowd had brought mini TV screens with them and were watching the eclipse on them, the main studio having been set up at St Michael's Mount, so we were able to huddle round and see a two inch TV version of the reality surrounding us.

At 10.45a.m., the temperature started to drop and the light become appreciably dimmer. Streetlights started to come on as their sensors recognised approaching night, and other sources such as traffic lights shone quite brightly. Camera flashes could be seen clearly from across the bay at The Mount. Very weird light effects were occurring in the cloud-ridden sky, but it still did not feel particularly special, just like an autumn twilight, albeit in the middle of the day.

Suddenly from 11.05a.m., the true magic took over. Light levels dropped dramatically, so that it was now very dark, like late dusk. All the birds suddenly flew away, and there was an eerie silence all around. People were visibly moved and huddled together in awe and, despite their knowledge of what was happening, fear.

11.09a.m.: Although we could not see the sun, events taking place around us were incredibly dramatic. Firstly, the band of brightness in the south and west was replaced by darkness, which rapidly advanced to cover the whole sky, with the feeling of a shadow rushing towards and over us, probably made even more disturbing by the fact that we couldn't actually see the moon covering the sun to confirm the cause with our own eyes. We didn't need to look at our watches to know it was now 11.11a.m.: the darkness was extreme.

People virtually cowered, and reeled around in shock although they knew it would only last two minutes. Then, suddenly the southwest strip was bright again, and within a very few seconds the whole sky seemed to be brighter than it had been all morning. Though possibly an illusion or caused by the ability of the eye to adapt to various levels of brightness, the light seemed to have returned much faster than it went.

Everyone looked relieved and much rejoicing took place. My own feelings were very high, having gone from a resigned acceptance that I would not be able to see the eclipse, through the dramatic experience of the sky and surroundings suddenly darkening and even more suddenly lightening again. A couple of people started to fly a kite with an eclipse pattern on it, saying 'that's as close as we are going to get', but I had found the whole event psychologically moving, almost disturbing, reaching down into very basic instincts. I was extremely pleased to have been there, and felt that the cloud-covered experience may have been better in some respects than watching an eclipse in a clear sky. Not knowing where the sun was, nor being able to actually see the moon covering it somehow made the feelings generated even stronger.

Eclipse

In the beginning *you* created all
Sun, Moon and Stars, all answered to *your* call
The Earth, *you* filled with life of ev'ry kind
With all its needs, abundantly to find
Resources of all types, free, everywhere
Water, food, heat, shelter, good clean air
All neatly balanced, according to *your* plan
But one became too clever / stupid: Man

Through ages they took over, lust to rule
All other life-forms answered to their call
They fought each other, taking sword in hand
Enslaving peoples, occupying land
All in the name of Justice, Truth and Right
Took what *you* gave *all* free, by force and might
Upset *your* balance, undermined *your* plan
No need for *you*, had clever, stupid Man

Almighty Father, time comes to show our worth
Let darkness spread across the tortured Earth
Turn off *your* heat and light for just one hour
Reveal to us how mighty is *your* power
Show what is to come, our dreadful fate
Unless we change, our time is running late
Restore *your* balance, give us back *your* plan
For we are not so clever, stupid Man

Mother Nature, Source of Life, we pray
Hear us calling, show again *your* way
Help us to see our role in Life is small
Just one of many, never 'best of all'
All forms of Life are equal in *your* eyes
You made them all, and equally *you* prize.
By being clever, ruining *your* plan
Accept us back, as simple, stupid Man

It is easy to see the incredible effects such events would have had on primitive peoples. Unlike us, they would have had no forewarning, nor knowledge of what effects could be expected or how long they would last. It is unlikely that any of them would have experienced or even heard of anything like it previously.

Their first indication that anything was happening may well have been the same as our first physical sign: to notice an unexplained fall in temperature in the middle of the day, their natural reaction being to look towards the sun. At first, with no smoked glass to look through, they probably would not have been able to see anything wrong, but would then have noticed the light beginning to fade. By the time it was dark enough for them to be able to see what was happening, most of the sun would have already been covered / missing (you don't notice the difference in light, and can't see directly that part of the sun is missing until a fairly advanced stage in the eclipse). They would have had no idea what could possibly be happening, other than that the sun, light and warmth were all disappearing before their very eyes. It is extremely unlikely that they would have immediately realised that the moon was involved, and if they had, that would have confused and frightened them even more.

For the next half an hour, they would have continued to watch as the world became darker, the temperature fell, and the sun just disappeared, apparently into a hole, or having been eaten. Imagine their fear! Then incredibly quickly the light would have returned, the temperature risen, and the sun start to reappear. With immense relief they would have watched as the sun regained its normal power and shape, and everything went back to normal. Except that, if what we were told about the dangers of looking directly at the sun is true, over the next few days many of their number would have gone blind.

No wonder such people put tremendous effort into studying the movements of the sun, moon and other heavenly bodies. The survivors would certainly have tried to look for an explanation, and may have realised that the moon had been getting closer to the sun in previous days. Imagine the power anyone able to predict such an event would have been able to wield.

With thoughts such as these, I found my way back to my tent. The lads had returned, having had much less success than I, and were cooking a consolation breakfast. Without Phil to drop his subtle hints: 'Cor, I would love a piece of bacon, but I haven't got any or a stove.' I thought I had had it.

'Have you got a plate then, John?' Brilliant. We tucked into chicken, sausages, bacon, eggs, tomatoes, mushrooms, beans, corn on the cob, tea, coffee and orange juice. Not a bad brunch, and just the kick start my diet needed for phase two of the walking part of the holiday. You may have noticed a slight diminution in the emphasis being placed on losing weight in recent days, but tomorrow, my birthday, would witness a restart, honest.

All the time I had been keeping an eye on the sky, and suddenly out popped the sun from behind the clouds. We were now able to watch the last half hour of the eclipse, through 'the best selling solar viewers on the planet' as it went from about one third to zero. Okay, it was intermittent with plenty of cloud still around; Okay, I missed seeing the full thing, but I had seen very strange effects and been deeply moved by the cloud-covered version. I had no way of knowing whether this was as impressive as a clear sky would have been, but I had been delighted with it, and was now able to see the partial eclipse first hand. All in all, a most satisfying outcome, and one I would cherish for a long time.

The rain showers continued, we washed up and chatted some more, then all decided to complete a lazy day by dozing in our tents. What's wrong with me? There are paths out there just itching for a boot.

Phil eventually arrived:

'Saw it all. Fantastic. Ninety-five percent. Clear blue skies. Everything.'

How does he do it?

'Wolves won 1-0, and I sat next to a lovely woman on the train back here. She works in London, but was coming home to Falmouth, works for a PR company…'

Me, jealous? Ridiculous. Anyway, I knew this was just another of Phil's fantasies about meeting a fantastic woman, which he claimed to do virtually every time he went anywhere alone.

Oddly, it never happened when anyone was there to witness.

'Even my blisters are getting better.'

Shut up, Phil!

He had decided the awkwardness and lack of sleep inherent in returning by car, along with uncertainty over seeing the eclipse, made the train trip preferable. A taxi driver friend had driven him to Reading station, and he had been able to keep a good watch on events initially at home, including the TV coverage, then en route, and finally at the station and on the train. He did confess, however, that the sky had not really gone dark, just like a very dull day, so I was able to convince myself even more definitely that I had had the best all round experience.

Phil had bought me a present: a new map case. 'Of course, if you don't like it, you can always just leave it somewhere. But this time, take the map out first.' What a wag.

We decided to wander into Penzance to see what was going on and have something to eat, so arranged to meet the Blackburn Boys in a pub Phil happened to know, The Star. We knew it was time to start getting back into walk mode, so had fish and chips, in a friendly traditional shop, where two slices of bread and butter and a mug of tea were included with the meal.

The Star had video screens all over the walls and above the bars, silently displaying a vast range of TV programmes and videos, not including the eclipse. It was very noisy, hot and crowded, which made a refreshing change, but did make it very hard to get served. Having fought my way to the front, I decided to get two rounds at once, so ordered four pints of Guinness. Someone behind mumbled something to his companion about how much it helped speed everything up when some idiot orders Guinness, and I gave him a bit of a look. He looked like he was going to faint, and quickly said 'I was only joking mate. Didn't mean anything. Sorry.' This was not the kind of reaction I was used to, but I suddenly remembered my new haircut and T-shirt. A skinhead nutter: not the person to cross in a pub. Perhaps the four-inch scar down my cheek and tattoo of death across my forehead also had something to do with it.

The Blackburn Boys arrived, and we stood chatting in the only space available, which was the doorway between two bars, with the consequence that we constantly had to move to allow people to squeeze past, which can be pleasure or pain depending on the squeezer. While carrying out one of these manoeuvres, Phil brought his elbow forward rapidly and knocked Arno's drink out of his hand. A good half pint of amber liquid moistened Arno's shirt and trousers, while the glass proceeded to descend in apparent slow motion towards the floor, where it very fortunately (or perhaps unfortunately, from Arno's viewpoint) hit his foot, which broke its fall so it did not smash, but slid across the floor where Tristan picked it up. Arno's reaction: 'Another pint anyone?' I liked Arno.

At chucking out / up time, we all headed back to camp. On the way, the lads invited us to join them in a trip to The Barn, which they had tracked down as the local disco, and was clearly where I had seen happy people returning from on our first Penzance night. It took place in a building, presumably ex barn, just down the road from the back entrance to our camp. It would have been rude not to join them. Arno needed to change his shirt, and I took the opportunity to put my black nylon see-through one on: sexy or what? What?

Possibly because everyone was celebrating the eclipse, the atmosphere was really special, and we were not 'the oldest swingers in town'. Arno appeared with a can of Bud for each of us, explaining, as if we didn't know, that draught might be watered down, and bottles break, so cans were the best option. He resisted adding that glasses can get knocked out of your hand and spill all their content down your shirt.

The main bar was elliptical, not ecliptical, and served both dance floors, each of which had their own DJ, and played different music throughout the evening, so there was plenty of choice as to which music to dance / listen to. Just the right number of people were there, so that you had to squeeze past to move around, but good progress could be made.

While settling in, we stood close to one of the dance floors, watching the gyrations, sipping our Buds and flexing our dance muscles. We suddenly realised we were all watching the same person, one of those ultra sexy women in an ultra sexy dress, doing moves with her ultra sexy body which make you feel strange, and which you have only seen before in your dreams.

Continuing the dream, I suddenly realised she was looking straight at me and using a long seductive finger to beckon me to join her. I was just getting my tongue back in my mouth and experimenting with placing one foot in front of the other, having forgotten how my legs worked, when Tristan pushed past and started cavorting with her. I really must get some new contact lenses. Still, at least I hadn't actually moved, so didn't have the embarrassment of being humiliated in front of my new friends by her giving me a look of contempt mixed with pity. So good to have a high self-image. Perhaps it *was* me she was beckoning, and Tristan with the eye problems, or he just muscled in on a good opportunity. I will never know. I have a very good idea, but don't actually know.

We all fell about, laughing with jealousy and marvelling at his good fortune – things like that just don't happen in real life – and what it was she saw in his skinny body and ugly, spotty face, compared to us handsome brutes. I asked Greavsie if Tristan was always that proficient at pulling, at which he laughed even louder and said he normally has trouble pulling a pint, and on one famous occasion had managed to gate-crash an Air Hostess' Hen Night only to end up watching them dance round their handbags.

By the time we had regained our normal composure, Tristan had failed to pass muster, or whatever he was supposed to pass, and Sex On Legs had moved on to her next victim. For the rest of the evening, one of the background entertainments was watching man after man sheepishly summoned onto the floor, writhed around for a few minutes, and discarded like a sub-standard spare part. I was determined to see what happened at the end of the night: would she finally settle on some lucky (?) guy, go home alone, or be smitten by a shy quiet man she found irresistible because he ignored her advances? Phil? As with most such determinations, I soon forgot all about it.

We all commiserated with Tristan and didn't make all that many comments regarding his staying power or chatting up skills. After mooching around, checking out both dance floors, all the bars, and the people, and supping another couple of cans each, we were 'up for it' and slowly progressed from minor hip movements on the edge of the floor to taking over the centre and having a laugh with all the other very relaxed and friendly revellers.

Once I start dancing, I tend not to stop, so wasn't really surprised when I realised the others had disappeared, and I was surrounded by strangers. Such situations can be embarrassing, but the atmosphere, density of bodies, and tracks being played were such that no partner was needed, so I just quietly bopped away enjoying myself. One of my favourite sayings is 'sometimes I sit and think, other times I just sit', and I was in a 'just dance' mind set.

I was literally jolted out of this 'let the music take you' state of semi-consciousness by two very presentable young ladies who started to dance with me, one pressing herself against me behind and doing a very enjoyable wiggle wriggle, the other dancing very close in front of me, with pelvic rolls and buttock shakes, arms wide apart and a 'come on' smile on her face. I felt like the meat in a sandwich and was beginning to make quality eye, and other contact, when who should butt in? Good old Phil, of course. 'Come on, we're all in the other room. It's miles better in there, I don't know why you are still here – no action.'

With anyone else I would have been pleased to see them, thinking that the balance of the sexes had been established and we could soon find a way of dividing four into twos, but Phil? Not good news. I made subtle gestures to him as to why I was still there, and that I wouldn't be offended if he went away. When they failed, I just tried to ignore him, but he is extremely thick skinned and hard to shake off. In any case, as soon as Phil had appeared and made it obvious he was with me, I had sensed a rapid loss of interest on the part of the girls, possibly neither wanting to risk being the one who sacrificed herself to some unkempt, unwashed, dishevelled, limping wretch so her friend could get off with a God-like creature – the latter being me.

I followed Phil into the other room, trying to look as though I was hard to get and dreaming that the girls would think I must be 'something special' and come after me, but knowing it was far more likely they would turn to each other and say 'That don't impress me much' with a sigh of relief. I complained to Arno how I had been getting off with two gorgeous ladies when Phil had ruined it, but he was totally lacking in sympathy and somewhat doubtful, the noise level making it hard to be certain exactly what he said, but something like 'what a load of ducking row locks.'

The rest of the evening was spent by us dancing 'round our handbags', except that as the style of music and number of people leaving made it obvious closing time was looming, I suddenly received a hard shove from behind. A girl pushed past, turned and did a few moves, whereupon I recognised her as the one who had wiggle wriggled behind me earlier. Now she pressed against me as she pushed across my front, did some very interesting body rolls and arm manoeuvres, turned her back and rubbed against me. I was starting to join in, when her friend appeared with a guy in tow, grabbed her and dragged her off towards the exit. Should I follow?

There was clearly something wrong with me that I even had to question it. Any red-blooded man… It must have been the eclipse having strange effects on me, because I didn't follow. She probably would have wanted her evil way with me anyway. 'All mouth and no trousers' strikes again.

I had a quick look round, but BBs and Phil were nowhere to be found, so I decided to nip into Tesco's to buy rolls, cheese and salad for tomorrow's breakfast and lunch, and a triple coffee fiddle. A crazy end to a crazy day, but at least I was still a virgin.

First and Last

My birthday!

So that you can see what a marvellous card and present I received from Phil, I have illustrated it in the box below:

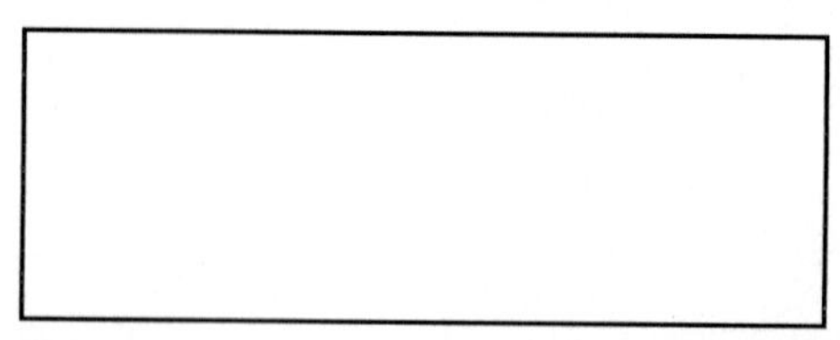

You can't see anything? Exactly. When you think of how I look after and say so many nice things about him.

You don't want to know which particular birthday I was celebrating, but I'll give you a clue: I'm now closer to sixty than twenty, as my son Peter took great delight in pointing out. His turn will come, and much faster than he realises. A couple of years ago I was in my twenties, now I'm thirty something plus, and it will seem like tomorrow when I start to draw my pension.

How would I be celebrating my birthday? Walking of course.

For the next few days we would continue to be based at the same campsite, travelling to and from the start / finish of our day's walk by bus or taxi. This meant we could leave most of our gear at the site, only taking the day's necessities with us. I'm not sure how it happened, but I had agreed to carry all of this in my rucksack, leaving Phil with only his camera to cope with. Consequently, I put both our spare jumpers, waterproofs, water bottles and food in my bag, which suddenly seemed to weigh almost as much as it ever had.

Their short eclipse break over, the Blackburn Boys were heading back, so we said our sad farewells, promising to mention them in the book, which I think they thought was nice. I hope they like it, especially stud, I mean Tristan.

That day, we would be starting from Hillside, near Geevor Mine, and intended to finish at Land's End. By the time we had packed, said goodbye, and walked yet again back to Penzance station / harbour complex, which included the bus terminus, we missed our bus (where's that fox) but decided to wait an hour for the next one rather than mess around with our itinerary even more by walking from Penzance to Lands End, i.e. the 'wrong way' for us, then filling in the gap the next day. We bought a coffee in the ubiquitous 'all day breakfast' café, served by the ubiquitous 'go slow' assistant. I have never understood the rationale of paying incredibly poor wages to incredibly poor staff. Surely a decent server would double the turnover, easily paying for any extra wages necessary to attract them, and result in satisfied customers.

As could have been predicted using Sod's law, when we were five minutes late for the first bus, it had left on time, but now we were an hour early, the next bus arrived twenty minutes late. For some inexplicable reason, large double-decker buses were used, despite the fact that there seemed to be relatively few customers, and much of the route was along narrow, overgrown, country lanes. Partly due to this, but also traffic jams in Penzance, we lost more time en route, and did not arrive at Hillside until almost noon.

As often happens, our late start, particularly as it followed a few days off, left us in a lazy frame of mind, and we decided to stop for a couple of pints in a pub after only walking half a mile. We then walked one more mile before deciding to tour Levant Engine House, where the 'man engine' has been restored to full working order, and was 'steamed up' every fifteen minutes.

'Man engines' were used to pull men, on ladders, up the mine shaft at the end of their shift, which saved them the two hour climb out, but on one disastrous occasion at this mine, had led to the death of most of the local men when the pit props had given way, and ladders, props and men had fallen thousands of feet back down into the black abyss.

The tour included viewing the remaining pitheads; where we were able to peer into the darkness and imagine what it must have been like to set off on your fourteen-hour shift every day.

Miner's Life

Two hours down the ladders starts my day
Two hours work before they start to pay
Two hours of darkness, aching, bathed in sweat
But can't afford to light my candles yet

I buy them from the owner: candle power!
Too high a price to light this ladder tower
I have to see the rock I'm working on
So hold the rail and trust no steps have gone

Make bloody sure I've everything I need
Candles, tools, helmet, blasting, and feed
Twelve hours to go before I end my stretch
Four up and down's too far to go and fetch

At last, I reach the current working thread
Crawl to the rock face, drill above my head
Insert the charge, retreat and off she goes
Clear the rock, drill on, repeat the dose

Now grab my lunch, I can't afford a break
Cornish rock pastie, rock drink, rock cake
Just four more hours until my shift is done
Then two dark hours climb out to miss the sun

By now it was mid afternoon, and a very hot summer's day, unlike yesterday, which further reinforced my belief in a slightly mischievous God. I had forgotten my hat and sunglasses, which was very bad news – as any serious walker could tell you, keeping your head cool in the heat and hot in the cold is an absolute must.

It was quite tough going round Cape Cornwall, and I soon started to fancy a spell in the cool dark recesses of one of the abandoned mines, but we were cheered by the sight of Longships, the rocks lying a mile or so off the coast of Land's End. The ups and downs continued to Whitesand Bay, where we walked along the back of the beaches towards Sennen Cove, dying for a cup of tea, an ice cream or anything else to quench our thirst, having finished our water bottles miles back. Numerous huts were scattered among the dunes leading up from the beaches to car parks, but they were all full of surf guards and their equipment, and not a single teashop or kiosk could be found. It seemed the only way we were likely to get a drink would be by pretending to drown, but we decided to keep that option as a desperate reserve.

As ever, the beaches were fantastic, with broad stretches of golden sand, fine sand dunes and glorious rollers pounding in. This time, however, instead of being virtually deserted, they were packed solid with rapidly reddening flesh. Phil claimed to have seen an advert for a beach umbrella which could be converted into a spit with a timer so the user could be slowly turned and evenly cooked all over. I knew he was joking, but could almost hear the flesh sizzling on some of the well-done bodies, and occasionally caught the whiff of roasting pork. I speculated as to what the reaction would be if I produced a carving knife and started to take a slice off someone's leg, but Phil thought they probably wouldn't see the funny side of it.

Finally, at the east end of Sennen Cove, our prayers were answered: a teashop. While Phil found some seats, I ordered a pot of tea for four, and persuaded the assistant to give me a glass of ice cold water which I soothed my parched lips and throat with while waiting for the tea. Normally, I treat water with the 'fish make love in it' attitude of the accomplished drinker, but this was pure heaven, and I started to think I was going to have a great birthday. I guess I could have given Phil some of the water.

The only seats available were on the sea-wall side of the path leading up from the beach, which would normally have been ideal as it gave a view of the beach area and the masses of humanity providing free entertainment while going through the basic chores involved in an English beach holiday. Presumably to provide shelter against potential wind and rain, however, the owners had very thoughtfully provided the area with a corrugated plastic roof and sides, turning it into a steam room on a hot day like this.

It was late afternoon, at the end of a baking hot summer's day, and a steady stream of lobsters were crawling their way up the path from the beach. Not only were they bright red and steaming, but they also held their arms away from their over-cooked sides so that they resembled lobster claws, and made little lobster sideways movements to avoid touching each other. While we relaxed over our tea we had a great time watching parents cope so coolly and calmly with their arm loads of sand covered beach gear (which turned it into first rate sand paper, just the job for tender flesh) and the fun and games little Joshua was having pulling Kirstie's hair, while Zack was refusing to go another step without an ice cream.

At the top of the path, to rub salt into sandy wounds, car park chaos was in full swing, with beach debris scattered everywhere while parents tried to load unwilling children into the back of steaming cars whose upholstery you could fry a leg on. Those slightly ahead in the game inched their cars out of parking spaces, expecting at any moment to puncture a tyre on an unseen umbrella spoke or chair leg, or at least run someone's kid over, only to join the queue waiting to block the exit. Naturally, this brought out the finer qualities in the drivers, each letting the other go first and accepting that they were as much to blame for the mess as anyone else, and it was all worthwhile as long as the kids had enjoyed themselves.

Phil, as you no doubt have guessed, knew of a *Good Beer Guide* pub in Sennen Cove, so it was not long before we headed for it, through the solid traffic jam along the single-track beach road. The Old Success Inn dates from 1691, used to be a smugglers inn, and overlooks a small but quaint harbour, where we admired the boats while quenching our thirst. From Sennen Cove to Land's End is one and a half miles up and along a gentle headland with marvellous views of the cove and out to sea, but even so, very few people were walking it, preferring the crowded beach below.

As we approached our goal, we could see what a mess this world famous landmark had become. The cliffs were badly eroded, and a theme park dominated the area, with a very plastic looking pirate sailing ship on the walls, hundreds of feet above sea level and looking ridiculous. Everything was very tacky, touristy and exploitative, including a charge of five pounds for having your photo taken with the Land's End sign. Everything was 'First and Last': First and Last Shop, selling First and Last T-shirts and First and Last cards, with a First and Last seat outside. I wished this could be my First and Last visit, though I knew I would be returning the next morning to continue the walk. The theme park consisted of moulded plastic 'olde worlde' buildings, with depressing shops full of expensive cheap tat, and arcade games. About the only thing I liked was a barrel organ, partly because I find the music very cheerful for twenty minutes or so, but also because it reminded me of a sketch from Monty Python in which a man had teeth, which danced in time to barrel organ music.

We bought a cup of tea in the large, holiday camp style, cafeteria and sat in the artificial square listening to the barrel organ. Twenty minutes having passed, the music began to grate, so we walked through the entrance gates and large car park to the bus stop.

While waiting for the bus, which true to form arrived thirty minutes late because we were early, we started to chat to a recently qualified South African teacher called Dee, who was spending a year in Bristol, using it as a base from which to tour the UK. Her plan had been to come to Cornwall for the eclipse with two friends, both had let her down but she had decided to come anyway. As you can imagine, this was all the encouragement Phil needed to launch into his 'best thing that could have happened' routine, while the bus wound its way round the country lanes back to Penzance and I dozed in the corner.

Dee accepted our offer for her to join us for dinner and a drink, and we agreed to meet at 8.30 p.m. at the bus terminus. It was already 7.45p.m. by the time we arrived in Penzance, so we rushed (Phil hobbled quickly) back to camp, whacked on loads of deodorant instead of having a shower and swapped our shorts for trousers to hide our dirty legs. Despite these desperate time saving measures, we were still ten minutes late and expected Dee to have given up on us if she had ever appeared, but as we turned the corner of the booking office, there she was.

By the time we had sorted ourselves out it was gone nine, and we discovered that eating at that ungodly hour was virtually impossible in a lively town like Penzance the day after its most important event for years. We tried several pubs, The Meadery, and even the fish and chip shop, but had to make do with a couple of bags of crisps and nuts: 'Good for your diet.' Thanks, Phil.

Over a few pints Dee told us how, having arrived late the previous evening, she'd had nowhere to sleep, eventually having to choose between dossing at the station or accepting an offer to share the pad of a dodgy guy she had met that day. She had tried to settle down at the station, but couldn't hack it, and had decided to take up the dodgy offer. She had been scared stiff, but eventually fell asleep with one hand on her money belt, and the other across her chest protecting other assets. She had awoken in the same position, then felt ashamed she had not trusted the guy who had been so generous to her.

During the day, she had found a room for that night only, so might return to Bristol the next day if she couldn't find somewhere for the following night. I offered her Phil's sleeping bag if she wanted to share our tent, and Phil didn't seem to mind, probably in the mistaken belief that he would be included in the package, but she said that while she might take up the offer of the tent share, she would buy her own bag.

We agreed to meet Dee at 8.30p.m. the next night, with or without her gear, depending on whether she had found somewhere decent to stay – not exactly flattering, but probably fair enough. The possibility of her not showing up due to either having set off for Bristol or, hard to imagine, having found better companionship, was an alternative scenario.

In my sleeping bag back at camp, I listened to 'John Denver' and reflected on a very different, but highly enjoyable birthday.

Spooky! Friday the 13th! I'm not superstitious (pass the salt) but we did have eighteen miles from Lands End to Penzance to cover, so I was rather wary that something could go wrong. Despite getting up at 7.00a.m. to catch the 8.00a.m. bus, we were still late and had to run up and over the hill leading to the bus station. The bus was nowhere to be seen, so we decided to 'wait and see', obtaining a coffee from Super Slick Service café in record time: three people in the queue, yet it only took seven minutes to get served.

Our tactic proved correct for once, and the bus arrived fifteen minutes late. Phil insisted on sitting in the front seat upstairs so that we could see the views, and promptly spilt his very hot coffee down himself, spending the rest of the journey with a large, steaming, wet patch on the front of his trousers. I helpfully explained to anyone coming up the stairs that he had just got a bit over excited at the fabulous views, and was normally fairly continent.

The views were, indeed, beautiful and spectacular, although the winding nature of the country roads, coupled with the desire to include every possible pick up / drop off point in the route, made progress painfully slow, a feeling heightened by constantly seeing signs saying 'Lands End eight miles'. Normally, I am very appreciative of attempts made to cater for the needs of the stranded walker, but now just wanted to crack on with those eighteen miles.

We set off from Lands End at 9.00a.m., and everything was perfect. It was too early for the fun fair or any casual visitors to break the enchantment, and we were able to appreciate the place at last (and first). There was a fine mist falling much of the time, which was ideal for keeping cool while trying to get some miles under our belt, and we began to motor. A steady flow of helicopters passed over our heads, plying their way between Penzance and the Scilly Isles, with numerous yachts, boats and tankers on a glassy sea. The two images were brought together at one point, when a jet flew past with a white trail streaming out of its four engines – like a speedboat in the sky.

Deserted mineshafts and buildings abounded, and there were dozens of beautiful, but inaccessible, beaches at the foot of sheer cliffs. Many of the mineshafts were topped by an elaborate metal cone-shaped grid, which we subsequently discovered was to allow bats to come and go freely. Gulls soared and hovered on the wind, as if to show who were really masters of their environment, and I must admit I would have loved to be able to copy them, but we both felt fit, strong and healthy, in fantastic weather, and bounced along the heather-strewn springy path.

The Minack Open Air Theatre at Porthcurno is a famous

landmark, including an amphitheatre overlooking the sea, but there was a charge of two pounds for a look around, which we decided not to pay on principle.

By the time we reached Penberth, eight miles into our hike, we were both more than half way through our water bottles, so when I saw two women standing outside a sea front cottage having a cup of coffee, I tried Phil's 'dying for a coffee, but haven't got a kettle' trick, only to be greeted with: 'No problem, there is a lovely coffee shop in Treen, one mile up the road'. Did they really not understand what I was saying? Did they think that an extra mile each way was a reasonable suggestion when they had a kettle within ten feet? Didn't they realise how much gratitude they would have received, or the scintillating conversation about the latest developments in blister technology they would miss?

At first we were just disappointed, but rapidly became dismayed, by the lack of refreshments anywhere. We ploughed on to Lamoran Cove, some twelve miles from our starting point, and the first place with refreshments available. It was 1.15p.m., only four and a quarter hours after leaving Land's End, which was fantastic progress given that much of the route was very tough going, with many ups and downs.

As is often the case, having pushed ourselves very hard due to the thought of all the miles ahead of us, we suddenly realised we had 'made it' in that we only had six more miles to cover that day. While this meant we could have relaxed and enjoyed the views more on the earlier parts of the walk, we were very pleased to be in such a commanding position. We celebrated with pots of tea and homemade bread and cakes, then set off again.

Pride comes before a fall, and it was only one and a half miles before we hit the bane of Phil's life: the dreaded tarmac, which proved to be almost constant for the remaining four and a half miles. Just to be sure we got the message, it started to pour with rain, which continued for the half mile into Mousehole (pronounced Mzl by locals) where I pretended to walk straight past The Ship Inn, but Phil was having none of it, just ignored me and went straight in. Two pints seemed to be on the bar immediately, so we assumed the barman had guessed what we would want, which looking back wasn't as brilliant as we thought at the time. One good pint deserves another, but then there was nothing for it except to plod, plod, plod down the tarmac roads into Newlyn, along the front and into Penzance. Phil's feet soon started to complain, shortly followed by Phil. You know the words by now (to the throbbing beat of 'The Sore Foot' march):

You don't know the pain I'm in
Okay for you to stand and grin
It's not fitness, hills or rain
But these blisters: oh the pain
Fine for you: don't care how far
All you get is mild BR

Each step I take my blisters rub
Let me stop; bring me a tub
Steaming water; feet inside
Gentle bathing, gently dried
Give me Moleskin or Compeed
And I'll soon be up to speed

Then we'll see who wins the race
Wipe that smile clean off your face
Run up the up, leap down the down
Under pressure, you won't clown
See you struggle, watch you crack
When I get my good feet back

Useless beggar, moaning git
Why do I put up with it?
I don't have it all my way
I felt a twinge – the other day
Pain is only in the mind
Other thoughts leave it behind

(Did you realise the last verse was not Phil, but me?)

Phil was feeling rather fed up, kept wanting to have stops, and began to lag, but I persuaded him to keep going as we were likely to stiffen up if we stopped for too long, and a long evening's rest would do us both good. I tried not to think about how I felt, concentrating instead on the sights and sounds around me. Newlyn had a large harbour, followed by a long sandy beach full of sandy people. A broad promenade, backed by the main road, went all the way to the harbour in Penzance, and I was impressed by the skills displayed by the good number of skateboarders and rollerbladers using it. We were also pleased to take advantage of occasional stretches of grass relieving the concrete / tarmac.

As we approached Penzance proper, we were surprised to see that only two people were swimming in the open air Empire Pool, and Phil was briefly cheered by the sight of the Isles of Scilly ferry terminal, which brought back memories of his free day, or day of freedom as he put it, and the other boats in Penzance harbour. For the ninety-third time we slogged up and over the road leading from the railway station to our camp, and finally arrived back at our tent at 5.30p.m., fully aware that we had done eighteen miles. Fantastic: three hours to relax and get ready for whatever the evening may bring.

Having flaked out for forty minutes, I felt sufficiently rested to go for a shower, which I knew would take a bit of effort but be well worth it. I started to get my gear together, when I noticed two very presentable young ladies starting to erect a tent close to where the Blackburn Boys had been, which had to be a good omen. I decided to play it cool, so just flashed a smile at one of them on my way to Shower Tower.

Depending on the time of day and cubicle you were in, taking a shower could be a very risky business. Sometimes the water was ridiculously hot, so much so that you had to balance on the edge of the shower tray to avoid a badly scalded foot, while attempting to flick a small amount of water out of the torrent flooding from the pipe onto your body. Similarly, it was largely potluck as to whether there was a hook or nail in the back of the cubicle door to hang your clothes on, failing which they had to be left in the washbasin area, where they might be knocked onto the sopping, muddy floor, or 'nicked in two minutes'. This situation also required positioning your towel carefully over the shower door so that it didn't slip off or get soaked by spray from the shower, and was available for use when needed.

I was relatively early tonight, and numbers were already down from the eclipse peak, which probably explained why the showers were particularly scalding, but I was at least able to choose a well-endowed door. I would have gratefully swapped for the opposite combination – no nail, but bearable water temperature.

By the time I had balanced and flicked to my heart's content, the new arrivals were just doing the finishing touches to their tent, so I felt it only right to congratulate them on a smooth erection, and we started to chat. It was not long before they told me they were from Denmark. I replied by saying that we had met loads of Dutch people on our walk, and it was a pleasure to meet some more, which gained a very surprised response:

'How do you know we are Dutch? I said Denmark.'

It was my turn to be surprised, because I hadn't even realised I had swapped countries. I had heard her say Denmark, and obviously knew Denmark was not in Holland, but somehow had registered them as Dutch, which it now seemed they were. Either I had picked up their accents or some other clue, or had become so used to meeting Dutch peeps that I assumed everyone was Dutch, despite all evidence to the contrary, and it was just a fluky coincidence that they were.

I told them we had met so many on this walk, I could tell a Dutchess in the dark by touch alone, if I ever got the chance, and the particularly interesting one, Berthe, gave me a look which said she might be prepared to test that theory to destruction.

They had arrived that afternoon, by train via London, and were going to start the coastal walk, going our way! This was the most convincing argument in favour of there being a God that I had heard in a long time.

Then I made my first mistake, in going to the tent for our maps and wonderful *South West Coastal Path 1999 Guide*. At first, this seemed to be a great move, since the girls grabbed the relevant map, spread it on the floor in front of me, and started to examine it in detail, which required them crawling around, and sometimes across, me. Suddenly, Phil appeared at the tent door, looking dreadful even for him, having crashed out on our return to camp and not moved since. Like a tramp on a bad hair day, who is wearing his worst set of gear, and suffering from a dreadful joint-stiffening disease.

'Lucky you woke me up' (the mistake) 'I was miles away.'

Then, to the girls:

'Heard you say you were going to do some of the South West Coastal Path, heading east. You don't want to go that way, the scenery is much better if you go from here towards Lands End, where we have just been. Did John tell you about my blisters? Of course, they're not as bad as they were, but look at this one under my right foot…'

What chance did I stand? Suddenly the atmosphere changed, not for the better, and within two minutes the girls decided to both go to the toilet. On their return, the less interesting one, Irene, announced 'we are very hungry and have decided to go for dinner straight away. It's a shame you aren't ready, or you could have come with us. Still, never mind, probably see you in the morning. Oh, and we have decided to go west as Phil suggested, so won't be going your way after all…'

Berthe just stood and looked me in the eye.

I had already mentioned the wonderful Coldstreamer Inn as a great place for us to have dinner together, and they had the cheek to say they were going there. I agreed to show them the route while nipping out to get two triple fiddle coffees from Tesco's. On the way, I mentioned that it was the first time I had known the sight of Phil's blisters make people hungry, and Berthe said 'That was just a lame excuse' Brilliant, crap jokes in a foreign language. Before I could reply, we had reached the parting of our ways and Irene stepped in, giving her a gentle but firm shove while turning towards me with a 'that don't impress me much' look which said it all.

I was so distracted by these events that I very nearly put salt instead of sugar in Phil's coffee. On my return, he was fast asleep, and I decided to leave him that way as long as possible, which proved to be 7.55p.m., when rain pattering on the tent brought him round.

'Oh dear, your coffee's cold.'

'Never mind, better go for my shower, we are meeting Dee in

half an hour.'

Despite his physical condition, Phil was ready inside twenty minutes, and we set off on the walk to Penzance Station, for a pleasant change. While Phil was showering, I had contemplated suggesting that we forget Dee, and follow the New Dutch to our favourite pub instead, but by the time he returned, had decided there really wasn't any point, the situation had been made perfectly clear, and it would only end in disaster. I also knew there was no point in commenting on Phil's latest performance, but couldn't help.

'I say, old chap. Don't you think you could have tried a little harder with those rather attractive ladies? They were as clear a godsend as that girl I met on Kilimanjaro Two. Just arrived, no ties, going our way, tent right next to ours. What do you do? Show your blisters and persuade them the views are much better heading west!'

I can't remember if I actually used the phrase 'I say, old chap', but I am sure you get my drift. Can you imagine Phil's reply?

'Oh, come on John, you didn't really want to spend the evening with those two did you? They were okay, but we have arranged to meet Dee. She is real class and has already been let down by two friends, imagine the effect if we do the same. As for the walking, it just wouldn't have been fair to let them come east with us – you must remember how fantastic the views were back towards Land's End, and we don't want to be held up by slow people. Anyway, you know what you are like – all mouth and no trousers. If you had got anywhere, you would have found some excuse for not coming up with the goods, like you always do.'

I don't know how close you were, but I was miles away, naively expecting some kind of apology. Still, I was right about one thing: there really had been no point in bringing the subject up. Don't want to be held up by slow people? What did he think he had been doing to me for the past week! All mouth and no trousers? If he didn't keep mucking it up for me…take that girl at the disco, doing great 'til he barged in…mind you, I did rather blow it at the end…perhaps he is right… I know he is right… I just keep pulling back at the last minute. What's wrong with me?

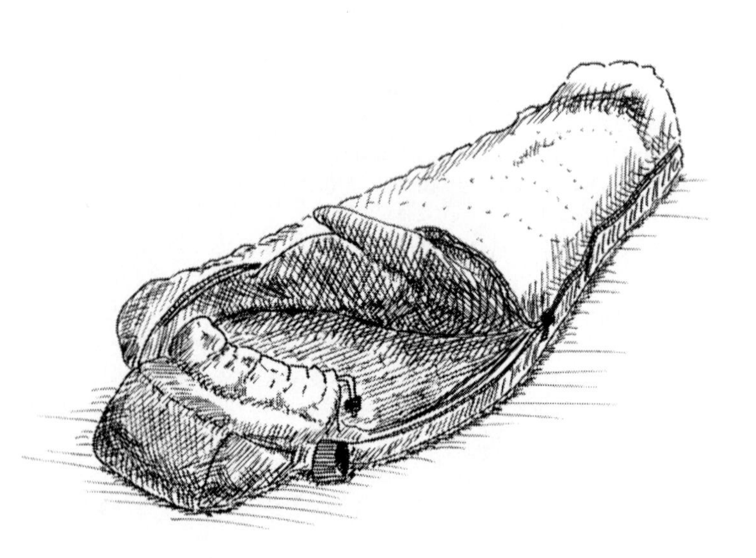

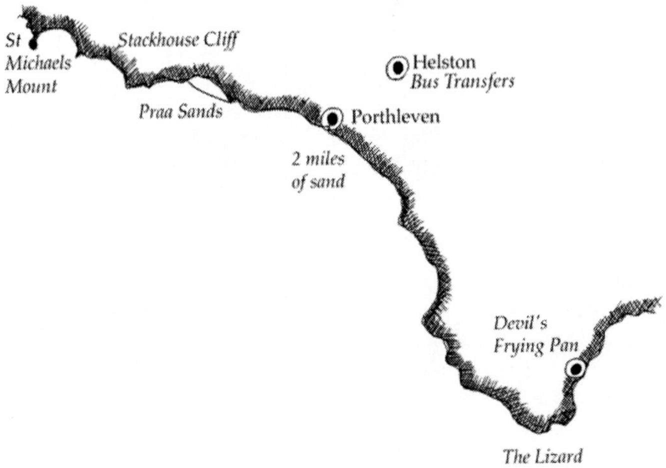

St
Michaels
Mount

Stackhouse Cliff

Praa Sands

Helston
Bus Transfers

Porthleven

2 miles
of sand

Devil's
Frying Pan

The Lizard

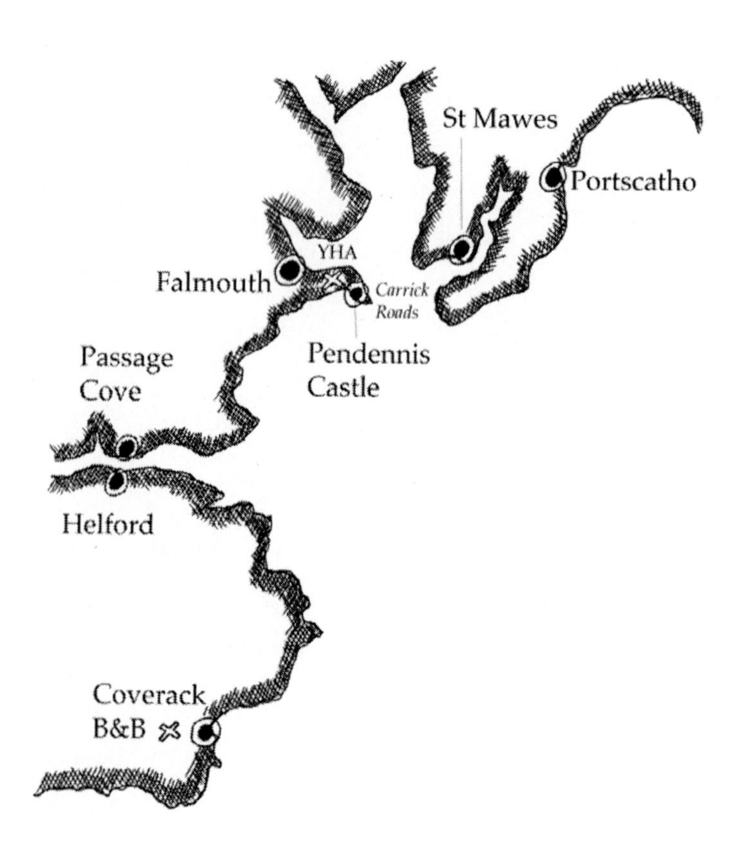
St Mawes
Portscatho
YHA
Falmouth
Carrick
Roads
Pendennis
Castle
Passage
Cove
Helford
Coverack
B&B

Unnaturally Warm

We met Dee as planned, but minus her gear – not as Phil had planned. All day he had been talking about her lovely auburn hair, fetching smile, captivating giggle, fascinating wiggle, but especially her beautiful, sad, grey/blue eyes, which revealed a dark secret of past torment. He had also said how great it would be if we could get some feminine company in the tent for a change, to which I replied he had the usual two chances of that, as seemed now to be confirmed. In any case, I found her sickly sweet, and had noted that she hadn't put her hand in her pocket all the previous evening – I didn't fancy subsidising a South African tour of the west country just so Phil could moon on about her. As for sharing the tent – it was already crowded with us two. I could imagine the chaos if she expected space to lie with one hand on each of her most prized possessions.

She told us she had found somewhere just for that night, so would like to put the tent offer on hold, possibly taking it up later. Following her fear in dodgy guy's pad a couple of nights earlier, I thought this was a valid stance, but Phil's disappointment was plain to see. The Meadery was very close to the station, we knew it didn't stop serving until nine, having arrived at ten past the previous night, and had seemed fine, based on a quick look round then. It was a large old warehouse, and had bars and restaurants on several levels, so we decided to try one decked out as though part of a ship, including a life-size pirate standing in the rigging, with a knife in his teeth and long, curved sword in one hand. The house speciality drink was, in line with its name, mead, which they claimed to be an aphrodisiac, so we could hardly refuse – a very clever marketing ploy.

Given the surroundings, we decided to really push the boat out. Dee was a veggie with a vengeance, and had plain salad. I had chicken salad, and Phil chicken and chips, the first time we hadn't had fish since – actually since the night before, when we had subsisted on bar snacks. All through the meal, Phil complained that his chicken had been hammered to death in the microwave, which was his way of saying he thought it was overcooked. I pointed out that it was hopefully already dead before being put in the microwave, it would be very difficult to hammer it while in there, and it may have been properly cooked on a grill.

Something was definitely being grilled. Phil was hot on the trail of Dee's past, and firing questions at a rate the pirate would have been highly impressed with had it been shells from a canon. She was loving the attention, seemed happy to reveal all, and I became quite fascinated in their dialogue and interactions, enjoying the role of voyeur.

By the time Dee had finished her salad, Phil had discovered she had been dissatisfied with a two and a half year relationship, so had suddenly upped and offed, and come to England six weeks previously. Her ex had clearly been distraught, because mutual friends had informed her he had already moved in with someone else.

Phil prides himself on his psychoanalytical skills, and was able to take these few facts and draw a profound diagnosis: 'He's blown it.' He then spent the dessert, coffee, bill paying and leaving sessions telling her how it was her who he really loved, he had just moved in with this other woman on the rebound, she didn't mean anything to him, he would take her back like a shot, she could get him back easily, but she was much too good for him, she had left him... (I'm just off to the loo)... obviously hadn't been right for her, she had made the right move, he should have followed her over here, now he stood no chance... (Hi, I'm back)... she would be getting all these new experiences and attitudes, she would outgrow him even more, he was just an idiot, stupid letting a fantastic girl like her go so easily... (More coffee you two?)... He was just trying to get her back the easy way by making her jealous, he couldn't be bothered to get off his arse and sort it out properly like a man, she was miles better off without him, she should be so proud of herself... (I'll just get the bill)... she had actually done something, she had made a decision and gone for it, she would never regret it, what do you think, John?

'Er, um, I was... er, I think it was better here despite the cloud.' Oops!

We agreed that, the night still being young even if Phil wasn't, we would try The Barn disco. Since this necessitated walking past our camp, we took advantage of the opportunity to show Dee the tent she could have shared, and still might be the following night, but didn't let her inside in case she had a delicate stomach.

It is rarely a good idea to attempt to replicate a great experience, and this was no exception. The Barn was just not the same place as before: there were fewer people and, possibly as a result, the atmosphere was nowhere near as lively and buzzing. We stood chatting, waiting for the Becks to take effect, and slowly relaxed into the new mood. Sure enough, the tracks played started to improve, and we found ourselves tapping, then rocking, to the rhythm, then gradually moving onto the dance floor. As usual, I let myself become increasingly taken by the music, and spent who knows how long in the middle of the floor completely oblivious to what was happening around me.

Ultimately, a second slow dance in a row was played, and I found myself surrounded by couples doing what couples do in dark discos when slow music is played. Very difficult to keep doing your own thing in such circumstances, so I turned to the others and…

Oh, they must be in the other room…

Must have missed them at the bar…

Perhaps they are both in the toilet…

Er, where else can they be?…

Oh…

Of course, they must have left.

I knew The Barn wasn't as good as before, but didn't think they were having that bad a time. They seemed to be enjoying it, and had hardly stopped talking to each other, which had been one reason why I had 'done my own thing'. Odd they should have decided to leave without telling me. It was not as though we were doing a long stretch next day, so they surely hadn't decided to make an early night of it and hit the sack.

I wasn't going to let their boring behaviour ruin my evening, so bought myself a Becks, and waited for the music to speed up again, whereupon I resumed my customary position and style. When the next batch of slow tunes came along, I decided it was time for Tesco's, had a triple tea for a change and bought tomorrow's breakfast, expecting to bump into Phil at any minute.

Phil was not in the tent, so once I had paid the customary visit to Shower Tower and he was still not around, I drank his coffee, wishing I had not put the sugar in.

When I started to get into my sleeping bag, I was very pleased to find it was nice and warm – usually it took a good five minutes to warm up. Suddenly, it dawned on me that for my bag to be warm, something must have warmed it up.

Oh my God!

He can't have?

They wouldn't have?

In *my* sleeping bag?

I found myself standing up, looking down very suspiciously at my hitherto wonderful, but now defiled bag, and feeling rather sick. This was even worse than the night I had woken to find Phil kneeling over my bag while attempting, with very limited success, to piss into a beer can because it was raining too hard to go to the toilet block. At least that had been on the outside – this was inside, where I put my naked, unprotected body.

Perhaps I am not the sharpest person in the world at picking up these things, but how was I supposed to know? There hadn't been any clues. And we are talking about Phil.

I took my bag over to ST for a close inspection under the bare light bulbs – suddenly everything seemed to be cold, hard and naked. A very thorough inspection inside and out put my mind somewhat at rest, but not completely – there were a few slightly suspicious stains, but none looked fresh, and at least I didn't find a used… which really would have been the last straw. Being careful not to touch any iffy parts, I gave my bag a very rough shake and extremely good wipe / sweep out with mounds of toilet paper. Even so, I decided I couldn't bring myself to get in it, so would sleep on top in my clothes that night, then either throw it away and buy a new one, or take it to the laundrette for a five star wash.

I lay awake and tried to rationalise the situation. My sleeping bag was tapered towards the leg end, making it a very snug fit for one, so surely too small for two, even if they were skinny runts like Phil and Dee. Perhaps I had imagined the unnatural warmth when I first got in, or had warmed it myself when sitting on it while getting ready for bed.

Needless to say, I had a restless and demoralising night, not helped by some thoughtless, or more likely deliberate, idiots loudly playing and singing along to music, which, naturally included Bob Dylan's 'lay lady lay, across my big brass bed'. They were put off when it started to rain at 3.47a.m., which I initially thought was great because I might get some sleep at last, but as the rain became heavier I became more depressed as I thought about the consequences.

Rain is always bad news for campers. It makes a loud drumming noise on the canvas, making sleep very difficult, and conjures up images of Flanders and Swan's:

mud, mud, glorious mud,
nothing quite like it for cooling the blood,
so follow me follow,
down to the hollow,
and there let us wallow
in glooooooooooooooooooooooorious mud.

Which is great if you are a hippo, no doubt, but not so much fun if trying to keep your gear clean and dry. Mud pools form around tent doors uncannily quickly, and it is impossible not to walk it inside, where it spreads everywhere and onto everything. Rain also acts as a precursor to a day's walking in uncomfortable, stuffy waterproofs, with constant risk of slipping over.

Sudden great thought: having wet feet is terrible for blisters!

Oh no – BR!

Although I was convinced I had been awake all night, I woke to find Phil fully clothed but soaking wet, with his boots still on and, whether by luck or judgement, sticking out of the main tent into the porch area. I decided to play it cool, so went to fetch a couple of Tesco Tripple Fiddles. By the time I returned Phil had… not moved a muscle, and I decided to let sleeping dog lie, so nipped off for a shower.

Soon after returning, curiosity overwhelmed me, so I

accidentally kicked Phil's legs a few times and dropped my shower bag on his stomach, which caused an involuntary twitch.

'Oh, you're awake at last. Here, sit up and drink this before it gets cold.'

'Ow… eh… oh… er… ta… oops!'

Phil slowly entered the conscious world and started to drink. After a couple of minutes struggling with myself not to say them, I suddenly heard the words the world had been holding its breath for echo round the silent tent.

'So, what happened to you last night?'

It was as though someone else had said them, and I felt detached, as if observing from above, though able to see and hear everything. Then I started to feel quite proud of myself – I could still hear the echo, and it seemed to me I had spoken very calmly and matter-of-fact, disguising my concern over my sleeping bag's honour. I looked at Phil to check his reaction, to find him obviously waiting for a reply. I had been right – someone else had said them – Phil.

My shock must have been obvious, since I almost fell on top of him, but he continued:

'We waited ages for you at the gate by the roundabout.'

'Wha…?'

'But eventually I just had to walk Dee home.'

For the next two hours, Phil tried to convince me that:

- He had brought Dee back to the tent, but only for a coffee;
- Neither he nor my bag had been involved in any night manoeuvres;
- She had sat on my bed, but any unnatural warmth inside must have seeped through – that's the trouble with those cheap bags – no insulation;
- How could I think he would do such a thing (what about when he peed in that beer can while holding it over my bed? And that was with me in it);
- He was back by three (I had been force-fed CD's until 3.47a.m. and he wasn't back then).

Ultimately, Phil had all the best lines and, I must admit with some justification, went on the offensive:

- You're not my mum, and I'm not fourteen (both hard to deny);
- You weren't interested in her (even truer);
- Okay, we did, but back in her B&B (fine, and congratulations in order, but how could I be sure no dry, or wet, runs in my sleeping bag?);
- If you don't believe me, buy a new bag (this was really mean – he knew I was an accountant);
- You're jealous because you blew it with that girl the other night, the barmaid, Birgide and Ingrid, Berthe, and every other woman we have ever met; (er… well yes but… erm…)
- Just because you are all mouth and no trousers, why should I reject a perfectly good opportunity to get my leg over? She was gagging for it, six weeks after splitting with a long-term partner – you know women miss it more than men;
- I thought you would be pleased for me.

This was a side of Phil I had never seen before, nor dreamed existed. In fact, it seemed to be a completely new Phil! He was strident. Forceful, relaxed and confident.

Dare I say 'cocky'!

I am extremely sensitive to my surroundings and fast on the uptake, and an incredible thought struck me. Surely he hadn't been a… I mean, he must have… it couldn't have been his first…

I started to feel the giggles coming on, rapidly worsened by my next thoughts – was Phil's voice suddenly deeper? Had his adolescent spots gone? Before laughing in his face, I managed to blurt out:

'Any limp problems?' and Phil shot (!) back with:

'Not until number three.'

We rolled around the tent in helpless laughter, whooping and yelping, giving each other 'high fives' and making all sorts of disgusting gestures, pausing only to clutch our aching ribs.

Eventually, I crawled out of the tent to find the two Dutch girls all packed up and ready to go, but sitting on their rucksacks, wiping tears of laughter from their eyes.

'Sounds like your friend had it good last night.'

'If I'd known you were all mouth and no trousers (said in a 'not sure what it means, but pretty good idea' voice) I would have let Berthe stay and play with you.'

'Do you think he would have blown it with me? Perhaps I would have blown it with him.'

'Cheek! How long have you two been listening in on our private conversations? Is nothing sacred on this site?'

'Not if you shout it at the top of your voice, mate' came from a tent further up the row.

They thanked us for the entertainment, and recommendation of The Coldstreamer Inn pub, which they had thoroughly enjoyed along with the company of two guys they had met there, who were heading west, in line with Phil's advice. They had decided to tag along, despite the attraction of sticking around to hear the next episode of our adventure.

By now, it was 11.30a.m., so we were already hours and miles behind our vague schedule. We bunged everything we could think of into my bag, donned our waterproofs and set off through the site, bowing and waving at the numerous people still chuckling away to themselves. We agreed that, although we had both walked it individually during our time at Penzance, it just wouldn't feel right if we caught the bus round the bay to St Michael's Mount / Marazion. Besides, two miles on flat, firm sand watching the tide come in and the Mount draw closer was a great way to start the day. It wasn't until we were halfway across the bay that I realised I had forgotten to switch from glasses to lenses, and had long trousers on instead of shorts, both being bad choices in rain.

Phil told me the full story of the previous night, but I will spare you the gruesome parts. He had been planning it all day, and much of the night before that, deliberately 'laying it on thick' with Dee, telling her how beautiful she was, what a fool her partner had been, any man would find her attractive etc., etc... He had known there would be no problem getting rid of me in the disco – 'just wait for you to adopt your "John Revolta" role.' He had sprayed deodorant around in the tent in case it had an odd smell – I told him there was no 'in case' about it – and confessed his plan had been to do the dreadful deed there, pushing our beds and bags together (aha, I knew it!) but Dee had invited him back to her place where she was sure they would be more comfortable. I set him up with 'she sounds like a wanton hussy' and he duly obliged: 'yep, she was sure wanton it'.

Phil was not in walking mood, and persuaded me to stop in the Godolphin Arms in Marazion overlooking the mount and harbour, for a couple of coffees while we read the paper. By the time we set off again, the rain had stopped and we were able to stow away our waterproofs. A mile of road walking was followed by a very pleasant stretch along a good path at the back of the rocks / beaches. We had lapsed into silence, not due to any animosity between us, but I guess because we both had plenty of thinking to do. Phil's were no doubt carnal, about the previous night, that night, nights to come… and how to get rid of the gooseberry. I was thinking 'What's going on?'

Since the eclipse, everything seemed to be very odd:

- Phil had pulled, highly successfully, a real, live, passable woman;
- His blisters seemed to be virtually cured;
- He had new bounce and drive;
- I really was all mouth and no trousers;
- A gooseberry!
- My back and legs were beginning to ache;
- I was starting to sneeze, my nose was running, and I felt shivery.

Perhaps that person I had seen near the triffid plants was a witch or druid, and had put a spell on me. Or Phil was sapping my strength, stamina, God-like sex appeal and basic good health, and would soon grind me into the dust while he went from strength to strength, striding mile after mile, and taking every woman in his path. Possibly a bit far fetched, but it did seem highly likely he would decide he had better things to do than spend another week walking with me.

Engrossed in such disturbing thoughts, I found myself at Perran Sands, with Phil pointing out the teashop and declaring it was time for a cuppa. As we started to drink, Phil said:

'John, about last night…'

I thought *Here we go. He is going to say 'I've decided to go back to South Africa with Dee for a life of debauchery and lion farming' or something* so was surprised to hear:

'As I was leaving, Dee said it had been fantastic fun and all that, but she had decided to go back to Bristol in the morning. Asked me to say goodbye to you for her.'

By the time he had finished, his voice was a bit shaky, so I said the kind words I thought he would want to hear:

'Just going for a pee,' and bolted.

I didn't know whether to laugh or cry, feeling sorry for Phil, but selfishly relieved to be rid of Dee since Phil would now carry on with the walk. After ten minutes, I thought it should be safe to return armed with a couple of 'all for the best' type lines, but Phil had the map spread across the table and spent the next ten minutes pointing out all the interesting places on our route, while I drank one cold and one hot cup of tea.

As we continued round the bay, we had a choice of either staying up on the headland as the book suggested, or down at beach level, so naturally ignored the book since we hadn't cocked anything up for a few days.

The beach soon began to be covered in rock outcrops, which became more closely packed and larger, with a flat slate-like structure, until we were engaged in full on scrambling along flat surfaces with frequent fissures and cracks, which were often too wide to jump across, forcing us to backtrack and find another route. We passed interestingly named features such as Favel's Hole and Long Zawn, by which point the rocks jutted into the sea, and were covered in slippery weed up to high water mark, which narrowed our choice. Ultimately, we came to Stackhouse Cove, where we were faced with drops of thirty feet or more on all sides, and after a couple of very dodgy attempts at climbing down realised we had done it again.

Going back would probably mean at least a mile before we could find a path up to the headland, and the outcrop now reached to within fifteen feet of the grass growing on the steeply sloping lower edge of Stackhouse Cliffs. Like the pair of completely stupid idiots we were, we decided that the next place name, Porth Sampson, was an omen, so agreed to risk our lives by scrambling to the top of the outcrop, climbing the fifteen feet of vertical rock face above it, then somehow making our way up the grass slope.

At just the wrong moment, a piece of rock came loose in Phil's hand, he lost his footing, didn't have a sufficiently good hold with his other hand to make a recovery, and came crashing down onto me knocking us both flying eighty feet down onto a flat slab below, to our deaths!

Obviously, we didn't die, but we so easily could have. In practice, we climbed to the top of the outcrop, using its many cracks and fissures as hand and foot holds. The vertical rock face was more problematical and frightening, but tackled in similar fashion, whereupon the final thirty feet of very steep and slippery grass proved to be, almost literally, our downfall. We kept very low and took it extremely carefully, but even so both lost our footing several times, thereby being totally reliant on very hastily grabbed handfuls of grass not pulling out of their tenuous rocky rootings. We got away with it, but I could so easily have been a dead man writing this.

We pressed on past Arch Zawn and Zawn Susan, assuming zawn to be Cornish for 'another idiot fell off here', round Cudden Point, passing Western, then Eastern Shag Rocks – which caused us both to smile without needing to say anything, and on to Praa Sands, where we celebrated our longevity with two pints of… Guinness, Phil befriending a massive Briard dog. As the name suggests, Praa Sands heralded the start of a long beach section, at the end of which we climbed onto a headland to make and eat our rolls, enjoying splendid views in good sunshine, and a stiff breeze – one of my favourite combinations, made even better by a blue sky dotted with light fluffy clouds.

Following a good, fast, fairly flat stretch passing several pleasant little coves, we found ourselves overlooking Porthleven harbour, just in time to miss the 5.24p.m. bus, which left little option but to have a pint in The Harbour Hotel, Phil keeping faith with Guinness, me trying the local Bosun.

It was now a beautiful late afternoon, and we sat on the harbour wall watching local kids having a great time jumping thirty feet into the water. On the bus back to Penzance, Phil was such fantastic company that I fell asleep; to be gently woken by him as we neared the Tesco roundabout. He had persuaded the driver to drop us there, and obtained a pocket timetable so we could better plan the next few days – I said he was fantastic company.

Phil said that, as I was obviously very tired and most of our clothes were badly in need of a wash, he would take them to the laundry in Penzance while I had a nap. I was surprised, this being the first time I had ever known Phil to give a two bob bit about clean clothes, but assumed he wanted some time alone to think over the Dee situation, so reluctantly agreed to stay there and sleep while he slogged up and over our favourite hill a couple of times. I tried to persuade him to take my sleeping bag, but he gave me a hurt look. While I packed the clothes into two bin liners, carefully tipped everything out of Phil's rucksack onto the floor, then stuffed the bin liners into it, Phil nipped across for a quick shower 'to freshen up'.

Having seen Phil on his way, I headed for the showers. No sooner had I whipped off my clothes than a woman came bursting through the door:

'Oh, are these mixed showers?'

'Dunno, but we can share if you like.'

'What's the point? I've heard you are all mouth and no trousers.'

I was virtually certain Phil's version of the previous night's activities was true, but while waiting for his return, decided to think it through. Phil and Dee had left the disco before me, and Phil had returned to the tent sometime between say five and eight in the morning. What they had got up to between times was none of my business, unless it involved my sleeping bag, but that was not going to stop me trying to puzzle it out.

Had they done the dreaded deed, and if so, where? Perhaps Phil had tried it on, been rejected, and reacted by telling Dee we were moving on. No, his reaction that morning and our subsequent discussions seemed to genuinely reflect someone who had definitely had 'a result'. They didn't come across as made up fantasies, unlike when he invariably claimed to have pulled when on his lone trips. But this success put those claims in a new light – perhaps it wasn't as rare an occurrence as I had always assumed, and his claimed previous conquests were genuine. Surely Phil couldn't be a globetrotting Casanova?

The most logical explanation seemed to be that they had done as Phil said, come back to our tent for a coffee and chat, things had developed, and Dee had made the very reasonable offer of a nice warm bed for the night. Waking up to find Phil next to her had sobered her up, and she had decided to call it a day. As Phil had said, with surprising insight, she had probably been missing the sex and warm feeling of a shared bed, but one night with Phil had taken the edge off her appetite.

Phil seemed to have realised and accepted the situation, which taken all round had to be a winner. I just hoped he hadn't got himself pregnant.

Having sorted out that tangle, I relaxed into a beautiful, peaceful sleep, to be rudely awoken by a bin liner full of nicely screwed clothes being slung onto my stomach.

We decided to return to our favourite pub for another helping of Moules Marinière (who said accountants are boring?), which we both followed with peppered steak. It was stiflingly hot inside, so we ate the starter outside, by which time it was cold and dark there, so we moved back inside for the main course, which was massive, cooked to perfection, and accompanied by fantastic crisp vegetables and roast potatoes.

Over dinner, Phil related the tale of the laundry trip:

'I arrived at seven-twenty, to find a notice stating 'last wash seven-thirty' – more fantastic luck – that fox of yours must be on overtime. Too much for one machine, so kept our stuff separate, like you said. Kept myself entertained reading the notices which included one on the tumble dryers stating 'no animals' which I found rather disconcerting, although I have heard rumours of cats in microwaves and one of my sick friends sent me an e-mail containing a 'frog in a liquidizer' game. As I was removing things from the dryer, neatly folding mine, I decided to strip off and change into my nice clean warm clothes, like the guy in the Wranglers advert, but Mrs Wash wouldn't let me, saying she was supposed to lock up ages ago, and Mr Wash would be wanting his supper.'

Back home, I decided to rearrange the tent layout to avoid a bump in the ground, which had been keeping me awake, but also so that our feet were 'downhill'. I always find it easier to sleep that way rather than either feet uphill or across the slope – I definitely didn't want to roll onto Phil, or vice versa, in the middle of the night – leave that to Dee. At first, I was kept awake by the guitar man, but he soon stopped, or perhaps I just fell asleep mid-verse.

I woke late, which is generally a good start to the day since it means I've slept well. In this case, however, it meant we had to run the last four hundred yards to catch the 9.15a.m. bus from Penzance. Since we intended to spend several more nights at the Penzance campsite, we decided to buy a three-day bus pass. Timetables also meant that the only realistic thing for us to do was make our way to The Lizard, via Helford, and walk west 'back' to Porthleven. We weren't dumb enough to think this would be 'breaking the rules' of our walk, and thought it would make a pleasant change to have the sea on our left, instead of right.

Even so, at Helford we still had to wait forty minutes for the bus to the Lizard. While I guarded our bags, Phil went to buy a paper, returning twenty minutes later to tell me about 'their lovely ginger cat called Marmalade. She sits on the same pile of papers all day – has done for fifteen years, which is very old for a cat. They also have a lovely bulldog – you can go and play with it if you like, I'll look after the bags.'

Having displayed one of his traits – love of animals – Phil now revealed a second, which I was much more in tune with. He spied The Blue Anchor, immediately recognising it as one of the country's first independent breweries, and declared that at least we knew where we would be spending the evening.

On handing our passes to the driver, we were informed they were not valid. A different company operates the Lizard Rambler:

'that don't impress me much.'

It was a forty-five minute journey, so we didn't reach Lizard until 11.30a.m., hardly the right time to be starting a long day's walk. Phil agreed: 'Let's have a cup of tea in that café.' Consequently, it was past midday before we made our way down the half-mile path to Housel Cove and turned right for Lizard Point. Being the most southerly point in England, with spectacular views, and a convenient two to three mile walk from the local village, we were not surprised to find the area quite crowded, but were soon on our own again as we made our way round Lizard and Predannack Downs. We found it generally good walking with occasional ups and downs to clear the lungs. My cough / cold was steadily worsening, and Phil was beginning to gloat that he had come through his tests, and now it was my turn. He was as supportive and reassuring to me as I had been to him about his blisters, constantly telling me: 'You're going down.'

We enjoyed many extremely good views, and the weather was improving, though still overcast with a cold wind. Neither of us fancied the rolls / salad I was lugging around, and Phil seemed to be deliberately hardly touching his water. When I broached the subject he laughed and confessed: 'you know one of your objectives for the walk is to lose weight. I'm thinking of your stomach: it must be shrinking. Time we had another 'fat photo'. Carrying my water is doing you good.' Thanks, Phil.

At Polurrian Cove, Mullion, we decided to stop for lunch. Phil, having slagged them off all week as being nothing but coloured fat, insisted on a cream tea. I reacted by being self righteous, and ordered quiche and salad – only to be told that it was not available on a Sunday. I gave in quickly: 'Oh well, cream tea then.'

Continuing west, towards Porthleven, we soon reached Perranporth Sands, which initially boosted our morale tremendously: 'Great, two miles on beautiful, flat sands and we are home' but the beach was, in fact, terrible for walking on, consisting as it did, not of real sand, but very small pebbles which gave as we pushed off every step. This meant we were putting in twice the normal effort, but only gaining half the normal ground. I soon had a very bad ache in my calf muscles, and was not a happy camper.

Two miles of flat sand seems to stretch away forever, and when progress is, literally painfully slow, the psychological effects are tremendous. This rapidly became my first real trial, and I had to resort to psychological ways of overcoming it. I looked for closer landmarks, which I could see were actually getting nearer, and would hopefully take my mind off the pain. I gave myself small targets to focus on, promising myself rewards if I achieved them – 'I'll rest after those rocks…when I do a thousand more steps… when I reach the path…' But as with the unspoken competition when climbing hills, neither of us wanted to be first to stop, so we kept plodding on. '…when I die.'

Eventually we reached Loe Bar, where Carminowe Creek and The Loe join the sea at high tide. The beauty of the broad river basin and surrounding woods helped distract my mind, but I was nevertheless extremely grateful to reach a stretch of fairly firm grassy sand dunes. All too soon, these ran out and it was back to more slog on pebbles, until we finally reached a lovely, firm path leading up from the beach. Heaven.

All the way along the awful pebbles I had dropped subtle hints for Phil to take a turn at carrying my rucksack, in which I had carried all our day gear the whole time we had been doing day walks from Penzance Base Camp.

'Of course, I am sinking much further than you because of the *extra weight I am carrying*'.

Nothing.

'I weigh three stones more than you, probably more like four and a half *including the rucksack*'.

'Yeah, must be tough.'

'It would be much easier if I didn't have *this rucksack… to carry*'.

Nothing.

I thought he must be getting his own back for me not lending him my sleeping mat when some bar steward nicked his 'tent', but wouldn't be surprised if he didn't even notice my hints: he is incredibly 'focused'.

Something I certainly noticed was a strange feeling in my right boot: a lump under the arch. Fearing that it must be a blister brought on by the pebbles, I decided to stop and check, to discover a large quantity of pebbles in both my boots. Having seen mine, Phil checked his, and sure enough, a fair proportion of Porthleven Sands duly appeared. How did they get in? Our boots were a very tight fit and we didn't seem to have been sinking sufficiently far in for the pebbles to have gone over the top.

My cold, and aching calf muscles had made me start to wonder whether Phil was right, and I was 'going down', but it was over a mile into Perrenporth on tarmac path, then road, and our roles were soon back to normal. There was a relieving sense of déjà vu as I strode ahead with Phil complaining in my ear 'these roads really muck your feet up'. Great, and to top it all, the sun came out. I had passed my test, and my spirits soared. Me going down? No way, Jose.

As we entered Perranporth harbour, we were tempted into having an ice cream at 'Nauté But Ice' then a pint at The Harbour. I noticed, for the first time, the cannons on the harbour wall, although they had presumably been there the previous day. Sunday seemed to be girls' jumping day since there were half a dozen taking turns to leap, screaming, the thirty feet into the water below, then climb out for another go. While this was probably far more entertaining, we found ourselves watching football (well, Motherwell v Rangers) on TV in The Green Parrot while we supped our beers.

Wandering outside, we noticed the irresistible smell of fish and chips on the breeze, so traced it to its source: Porthleven Fish Restaurant, where we each bought a take-away and returned to the harbour front to eat it. Porthleven Brass Band were just finishing 'Mamma Mia', whereupon the compère announced 'The band will play "Amazing Grace", and then we will have some music'. A ripple of laughter made her realise the implication, so she used the 'just seeing if you are awake' get-out and all was fine.

The atmosphere was warm, friendly and inviting, and I was just thinking of suggesting we forget the 7.45p.m. last bus, listen to the band and choir (obviously the compère's highlight) then catch a taxi back to camp, when the bus appeared seven minutes early, stopped for five seconds, then started to pull away. Phil moved at lightning speed, ran across the road, forcing a cyclist to swerve into a couple of people watching the band, and banged on the window. The driver stopped and opened the doors with only the faintest trace of a sheepish grin on his face. With no time to state my case for staying put, we clambered on, with several other grateful passengers, and off we went, still five minutes before time. I think buses and trains that leave early, particularly if the last of the day, should be shot, and so should the drivers. I decided to forgive him, however, because we had caught the bus, he had let us on with our half-eaten fish and chips, but mainly because I was too tired to bother.

We arrived in Helston, and commenced a skinful night in the Blue Anchor. It is a very strange fact that, no matter how tired I am, I can always summon the energy to lift a pint to my lips. Phil was now able to supplement his encyclopaedic knowledge of *The Good Beer Guide* by cross-questioning the barmaid, so we soon knew that the Blue Anchor is a fifteenth century thatched pub which was once a monk's resting place and later a tin mine's pay office, before becoming a 'micro-brewery' pub two hundred years before the current fad. Responding to our interest, she gave us a micro tour of the microbrewery, introducing us to the not-so-micro owner. They have strong and light beers, but their middle one is called 'middle' so that was easy to remember, and drink. As we started to convert our third pint of middle into piddle, I suggested a toast to Dee, which Phil readily agreed upon, but gave no further revelations as to what the true situation was between them. I'd have to be more cunning.

At some stage during the evening, as tends to happen on a skinful night, I started to feel a little homesick and guilty, so decided to phone my other half: 'Ring me back some other time, I'm watching TV now. Oh, by the way, I won't be coming to meet you. Bye.' It's nice to feel wanted.

Several pints later, we realised we had missed the last bus home, so persuaded the barmaid to ring us a taxi. Taxi drivers always have a story to tell, and this one was no exception. On discovering we were from London, he told us how he was mugged in a Soho strip joint in 1974, when he and some pals on a lads' night out in London paid extra for 'a special' but having taken their money the rather large men at the door had told them to 'piss off'. I thought better of suggesting that he really should have come to terms with it by now. Next, he told us that he takes every opportunity to pick up fares like us, who won't be sick in his car, rather than the local drunks from the disco. Poor innocent fool – still hasn't learnt from his 1974 experience.

I slept quite well until 3.00 a.m. when Phil attempted to slip out quietly, an event guaranteed to wake even the soundest sleeper. On seeing I was awake, he said he was 'just nipping out to the John, John' so I joined him, partly just to be sociable, but mainly because I knew that, given all the beer I had supped, having been woken I would soon want to go anyway. Once out of the tent, I was very pleased to be up, because it was one of those fantastic night skies in which the stars seem to be almost touchable, inviting you to collect a handful for a sparkling necklace to brighten your dark times. The Milky Way looked like wispy smoke blowing across the sky from a vast fire on the other side of the ocean.

I was reminded of a night spent in Death Valley, when one of the park rangers gave a talk about the constellations, signs of the Zodiac, and beliefs of American Indians, Romans, Greeks, and Norse, while a couple of hundred spell-bound campers lay flat on their backs, staring up at the heavens, thirsty for knowledge and eager to believe.

Now, I could easily see The Plough (Big Dipper or Great Bear, Ursa Major), Orion's Belt, and parts of Scorpio. If I remembered the story correctly, an Egyptian goddess had placed a scorpion into the sky to chase the hunter Orion – what better creature to use? As Scorpio rises, Orion dips below the horizon, thereby showing the validity of the story.

As had been the case with the eclipse a few days before, I started to think about the fantastic influence the heavenly bodies must have had on the lives of ancient people, and the power they gave to those knowledgeable of their movements. I had been in Death Valley when a comet passed close to the earth, looking like a light shining through thick fog. To ignorant people, it may have resembled a burning torch with smoke streaming across the night sky, as the Milky Way did to me now.

I like to think that even the most primitive people actually had a fair idea of such basic facts as that the earth is spherical. It seems incredible that sailors, for example, did not realise, and much evidence exists that many ancient religions and civilisations such as Mayan and Chinese knew that the earth was a sphere, and moved round a spherical sun. It really only seems to be western religions like Christianity that insisted the earth was flat and the centre of everything, and maintained their belief by murdering anyone who refused to accept it! When Columbus sailed the ocean blue, in fourteen hundred and ninety-two, I am sure he knew pretty well where he was going, even if he did have to pretend otherwise for the church.

I guess the modern day equivalent of Galileo trying to reconcile his scientific observations with contemporary religious teaching is the conflict between evolution and creation theories for the origin of humanity. Many people currently seem to believe both, reconciling between them by saying that the Adam and Eve story in the bible is an allegory, and completely glossing over such issues as, if we evolved we surely can't have souls, and there cannot be an afterlife.

You can probably understand that, with thoughts like these in my head, I had little sleep for the rest of the night.

Oh, the other interesting thing was that it took Phil four hours to return from the loo!

So Long, And Thanks For All The Fish

I awoke with a fairly good cough settling in, which made me really want to spend the day walking.

In order to continue following our tactic of catching buses to / from day's walks, we needed the bus from Penzance to Helston, so as to catch the Lizard Rambler bus to Coverack, find a B&B, walk to the Lizard, bus back to Coverack, and have a relaxing evening. Rather complicated, but a plan at last – properly thought out, and checked to bus and tide tables: now to see if it worked.

I decided to get two coffees, some breakfast stuff and cough jollop, from Tesco's, and Phil tagged along. Typically, there being two of us, it took twice as long as it would have for one, since we discussed what to get. We also, as would be expected, ended up buying loads more breakfast stuff than either alone would have, then agreed to buy another mound for lunch. Having bought and paid for all this, we spied the 'breakfast specials' board, with its irresistible offers. Being a natural 'Slenda Glenda' I went for 'The Magnificent Seven' (sausage, bacon, tomato, egg, beans, toast and hash browns) while Phil restricted himself to 'The Fab Five' (same but minus tomato and sausage), swapping the beans for tomato.

Having the breakfast specials was, very obviously, a bad idea, and showed how ridiculous impulse buying can be. Firstly, it made a complete nonsense of all the breakfast stuff we had just bought, including a full litre of milk to have with muesli bought several days previously. Secondly, we were finally moving on from our wonderful Penzance base that very morning, if we ever got round to it, so anything we didn't eat now would have to be carried with us. Thirdly, it took up too much time, partly due to slow service by another couple of those people who seem to be living their lives at half speed, yet are for some unfathomable reason highly prized by 'fast' food outlets.

We were supposedly aiming for the 9.00a.m. bus from Penzance, a mile's walk over heartbreak hill, but didn't leave Tesco's until 8.20a.m., getting back to camp at 8.25a.m. Ignoring reality, we rushed to pack, throwing away various items accumulated over our nine nights in Penzance, which were now either not needed or too much bother to take with us. We were living illustrations of the proverb 'more haste, less speed' as several rushed attempts at packing sleeping bags and clothes failed, culminating in our hurriedly folding the tent, which was damp on top, and dripping with dew underneath, and did not roll up as small as it should. With the laughter and good wishes of the site owner and several campers ringing in our ears, carrying several bags each in addition to our rucksacks, we set out at just after 9.00a.m., pathetically hoping the bus would stop for us at Tesco's roundabout if we flagged it down. It did, so our good planning had worked after all.

By the time we reached Helston, we had condensed everything into two bags each, having occupied most of the floor of the bus while getting sorted, so were feeling rather pleased with ourselves, instead of recognising what a pair of steaming wazzocks we had been. It had been more fun our way, though, and had given a lot of people something to brighten their day.

Transfer from the Penzance bus to the Lizard Rambler took place in the car park of... Helston Tesco's. We had just enough time for a coffee – unfortunately the cashier had not been informed of the special Triple Fiddle arrangement – before spending an hour finding out how the Lizard rambler earned its name, as it took us on a wandering tour of the Lizard area before arriving at Coverack, less than twelve miles from Helston. Phil told me the trip was 'tops', and I took his word for it, having spent most of the time catching up on the beauty sleep I had missed the previous night. It seemed that Phil didn't need to catch up on the sleep he'd missed while doing whatever he had been doing between 3.00a.m. and 7.00a.m.

Coverack proved to be a very picturesque village, with a nice harbour and beach, surrounded by cliffs and rolling hills. After several knock-backs, we manage to get a B&B, which seemed ideal, but was only available for one night, when we wanted two. The owner, who was busy painting the walls of a two room extension he had built just too late for the eclipse rush that didn't happen, agreed to get his wife to fix us, which we assumed was a good thing.

We dumped all our gear, apart from my daypack, and set off, soon reaching Treleaver headland where we had good views ahead of virtually the whole route to The Lizard. I pointed out the increasingly strong wind, which Phil blamed on the cooked breakfast. A couple of downs and ups led to Kennack Sands, with the very interesting Caerverracks rock jutting out to sea. We decided to have our lunch overlooking the beach, before descending to take advantage of the teashop. Phil spent the time there indulging in two of his favourite endearing little habits, firstly loudly pointing out all the fat women and secondly becoming temporarily stuck on a certain word or phrase, repeating it so often that I end up having to get away from him before I push him off a cliff or under a bus, depending on the circumstances.

This time, as it often is, the word was 'amazing'. Everything was amazing. There was a teashop but no lifeguard – *amazing*. All these fat women – *amazing*. Pot of tea one pound eighty – *amazing*; those cards, this beach ball, cream teas – *amazing*. Look at the view – *amazing*.

'Phil, for pity's sake, shut up for five minutes.'

'Can't believe you just said that. Amazing.'

By the time we set off again, it was 3.00p.m., leaving us one and three quarter hours to cover five miles if we were to catch the last bus back from Lizard, which should not cause a real problem providing we didn't allow ourselves to get distracted. On spying the pub in Cadgwith Cove, suitably named Cadgwith Cove Inn, we stopped for a pint.

It was very welcome, but left us extremely short of time, which, I was beginning to realise, was how we liked it. We both enjoyed the extra thrill of living on the edge – even if catching the bus wasn't a real life or death situation. As if to emphasise the point, Phil insisted on taking a photo as we left, in a very 'devil-may-care' manner.

Phil was on top form and thoroughly enjoying himself, and why not? We were having a relatively easy day, walking in an excellent part of the world, in perfect walking weather, had just had a pint in a lovely old inn, were facing a final challenging push to the day's finish – and I had a steadily worsening cough. It was easy to see how everything was *amazing* from Phil's viewpoint, and he wasn't having to listen to some idiot constantly telling him how *amazing* it all was. He also had the pleasure of being able to point out to me how my cough was progressing *amazingly* well, periodically repeating 'You're going down'.

I was confident I could fight it off and still leave Phil in my wake, but was struggling to ignore them both. Normally I enjoy the banter between us, and could hardly complain given the stick I gave him over his blisters, but I was finding him extremely painful just then. *Amazing*.

Having more or less deliberately placed ourselves in severe time trouble, Sod's Law took effect, and we came upon one of the best sights of the whole walk. The Devils Frying Pan, a deep pool about two hundred yards across, surrounded by towering cliffs jutting out from the surrounding headland. The only gap in the cliffs is on the seaward side, presumably having been worn away and undermined by constant wave motion, so that now there is a fantastic natural arch completing the circle. Phil took one look: '*Amazing*.'

Luckily, he was several yards behind and downhill of me when he said this, probably fiftieth repetition inside half an hour, otherwise I might well have succumbed to the Devil's temptation by picking him up and throwing him far out into the frying pan. I was not going to give up hard earned height just to murder Phil, so decided to punish him instead by racing for the top of this very steep climb.

It was during that ascent above the Devil's Frying Pan that I had my strongest sensation of Phil being a devil on my back. I came steaming and puffing, wheezing and coughing, up the slope, like *The King* on a very bad climb, on a very bad day, but could hear Phil just behind all the way, lightly leaping from rock to rock, like a mountain goat. At the top, I turned to find him looking cool and relaxed, not even panting, with an evil glint in his red, slit eye, a pitchfork in his right hand and a four-foot tail poking proud and upright from just above his arse.

While I stood and stared, gulping for oxygen, pulse racing and pounding in my ears, he calmly said:

'Good to stretch the legs and get some air in the lungs for a change, shame it was such a short up though. No real challenge' and set off, with me trailing along behind, a broken man. I don't think he had even noticed the state I was in, which is probably just as well, since if he had said 'you're going down. *Amazing*' then, I probably would have cheerfully thrown myself into the Devil's Frying Pan.

We decided to go via Cross Common to avoid the detour round the Lizard lighthouse headland, which we had seen the previous day but had intended to have another look at. A final very brisk walk up half a mile of road and we hit the bus stop three minutes early, to cater for an over zealous driver, as at Helston. Cornish ice creams had just been devoured when the bus arrived, the driver agreeing to arrange our transfer to the Coverack bus, phoning the other driver to ensure she waited at the rendezvous – just like the service in London.

In Coverack, we returned to our B&B and took it in turns to doze on our beds and soak in the freestanding, large, very stylish bath. It was supported by two substantial planks of wood, which required quite a step up when getting in, but made it feel very special, almost regal. Two old fashioned gold taps, and an abundance of massive, fluffy towels, completed the feeling of luxury, while the long, sloping end was absolute heaven to lie back against – incredibly relaxing. I soaked up the pleasure, and began to drift into that state of low-level awareness I call 'Mind Morphing', where the mind starts to wander, playing with recent events and future possibilities. I once, having never touched any drugs in my life, honest your honour, wrote a poem to describe the sensation, and danger of going too far. I don't understand half of it, which my son, Peter, says is a very good sign, but won't tell me of what:

Mind Morphing on the fine line
Look okay and feeling fine
Thoughts go wandering behind, ahead
Awake, asleep, alive or dead
Freely dreaming loose control
Imagination starts to roll
Hearing light and seeing sound

Tumbling, twisting round and round
Climbing up but falling down
Crying laughs with a smiling frown
Shouting tastes I never felt
Into another realm I melt

Mind morphing on the fine line
Looking rough, but feeling fine
Experiment with mental flow
Close to the line, I need to know
Where are the limits of this trip
How near the drop without the slip
How close is madness, where's the edge
Is genius on a mind leap's ledge?
Where is the boundary to my thought?
Is random thinking dangerous sport?
Pain or pleasure, dark or light,
Pleasant dreams or nightmare fight?

Mind morphing on the fine line
Look like hell, but feeling fine
Fighting pressures to conform
Suppress my dreams and hit the norm
Taking time for thoughts to roam
Fearing if I go back home
I might not find the flows again
To drift around a floating brain
Swimming pools of shining thought
Almost touched yet never caught
Don't want to chain or check my mind
Just set it free and drift behind

Mind morphing on the fine line
Total wreck, but feeling fine
Slow withdrawal from my jump
Then waking with a fearful dump
Back to face reality
'Want a fag or cup of tea?'

Pay the price and make a quid
Get a job, house, partner, kid
When they ask me what I've done
'Mucked' it up, or had my fun
Tell them I was feeling fine
Mind morphing on the fine line

Was Phil the Devil, or some fairly close relative? He seemed to be totally unaffected by the climbs, lack of sleep, and weight we were carrying – I was carrying. Had I imagined the red slit eyes, tail and cloven hooves at the top of Devil's Frying Pan?

What had Phil been doing the previous night which made him take four hours to go to the loo? I had assumed he had wandered off to think about Dee, but had he actually been doing devil deeds, or performing some satanic rituals – perhaps that explained the dead dog we had seen near Tesco's roundabout that morning – killed by Phil while enacting some sacrificial rite. Oh no – I had only taken a quick look, and assumed it was a dog – was it actually my lucky fox, captured by Phil so he could turn its good luck bad against me?

He liked people to call him The Cat – they are supposed to be lucky, but we had had a devilish amount of luck on this walk. I had been crediting it to my lucky fox, but was it really due to a black cat? Did Phil kill lucky animals and somehow take their luck for himself? By drinking their blood?

What had he really been doing during the eclipse? How did I know he had really gone back to London for a football match? How ridiculous was that? Far more likely he had been leading the druids at Stonehenge.

But what about his terrible blisters during the first week? He had seemed to be in incredible pain, I had seen them and they certainly looked genuine and terrible. Then again, how did he recover from them so quickly?

Aah! I see! Some crucifixes must have got into his boots and burnt his feet when he put them on. He had removed them somehow, then taken a week to recover, or performed some kind of ritual while at Stonehenge. He'd seemed very keen to get out of Cross Common that afternoon, and our inland short cut had avoided Church Cove. It was all becoming so obvious.

But if he had all these special powers, how could his problems on Kilimanjaro, and long-term suffering from glaucoma be explained?

Then the final piece fell into place – he had been so disgusted at failing Kili, and in such pain from the galloping gut rot, that he had made a pact with the devil. In return for perfect health, super-human powers and fitness, and being irresistible to women (named Dee, recently arrived from South Africa, and suffering from a recent break-up) he had agreed to…what?

Sacrifice me!

Grind me down.

Mug me into a fatal accident.

That was why he had insisted on climbing up at Stackhouse Cliffs / Porth Sampson. When that had failed, he had waited until the Devil's Frying Pan, then driven me mad by constantly saying 'amazing' hoping I would crack, rush at him, and when he dodged, fall to my death on the jagged rocks edging the fathomless, dark pool. Wait a minute! Sampson had been betrayed by Delilah – what was Dee's full name? Was Dee in league with him? Or had he cast some dreadful spell so he could have his evil way with her? Dee and Phil. Dee and Phil? DeePhil – Devil!

Obviously I knew it was all rubbish, and I was just having a few minutes mind morphing, but this was rather spooky, and I was really getting myself scared now.

Suddenly the bathroom door flew open and Phil stood there, minus tail:

'Hurry up, John, you've been in there for ages. I want to discuss my plans for tomorrow with you.'

'(gulp)…plans?…for me?…wh… what plans?'

'Well, we have rather been relying on luck, haven't we? I think it's time we started to really plan ahead, and stick to it, particularly for the next couple of days when we have to allow for bus times and tides at several river crossings. I don't want anything to go wrong. Sooner or later your luck will run out and your guardian angel won't be around to look after you.'

Having consulted maps, guide, tide charts and bus timetables, we decided that the next morning we would catch the 9.20a.m. bus to Helford, then walk back to Coverack, taking our time, but hopefully returning early enough to arrange somewhere to sleep that night if Mrs B&B hadn't fixed us.

That evening, West Ham were playing Aston Villa live on Sky, so I wasted half an hour looking for somewhere in Coverack that was showing it. Eventually, I accepted defeat and we spent the night supping Guinness in the Paris Inn, and eating steak and salad (me) and scampi and chips (Phil). I pointed out that he had spent the last two weeks slagging off everyone who ate chips, but he said that was different – they had a lousy lifestyle.

In the pub we met some people staying at the youth hostel, so on the way back to our B&B, we booked in there for the next night, relieved to find they had a very generous definition of 'youth'. We now had the peace of mind of knowing we had a roof over our heads for two nights running. The thought of getting the tent out at the end of a day's walk definitely did not appeal after our lazy time in Penzance.

In bed, I decided it was time for a progress review. For the first week, we had set ourselves very tough walk-oriented targets, and although the route proved much harder than expected, I felt we would have stuck to them but for Phil's feet. We had also been extremely casual about planning, relying on brute force and ignorance, with massive dollops of luck to pull us through.

In reality, while we blamed Phil's blisters on bad luck, they were evidence of poor planning and decision making when we pushed too far on day one. I could shift the blame onto Phil by saying that he should have said earlier that he had blisters starting, but that wouldn't be fair. I'm a great believer in teams sharing responsibility – it's just that, if he had said earlier that he had blisters starting...

During the last few days, we had completely changed our approach to include enjoyment, relaxation, coffee stops, Guinness, etc., and using B&Bs and buses rather than carrying all the gear. At long last, we were also becoming more focused on real planning, and had vowed to pay more attention to carrying them out.

The extent to which we were failures was a matter of opinion.

Taken from the hard line viewpoint, it could be said we had moved completely away from our original intention of a three-week hard slog, and were now planning much easier, softer targets. In the short term, no doubt the walk would be more enjoyable, but would we look back in years to come and think we had wimped out?

Were we becoming more mature and holistic in our targets, or turning into two old softies? That made me think about my personal target of losing weight. I felt a little slimmer, and had tightened my belt, but perhaps that had stretched – only the fat photos would really show the truth.

Until that morning, I had felt completely confident that I would be able to keep going, and had found the pace, slowed as we had been by Phil's blisters, quite comfortable, but Phil's complete recovery and my worsening cough had sewn the first seeds of doubt. I knew I could always insist on Phil carrying more, but that would be an admission of weakness in itself. If the truth were known, with his new eye drops Phil was probably fitter than me, but I didn't want to admit it so would continue acting like a donkey!

Extremely loud screeching and squawking by squabbling gulls on the roof above my bedroom woke me at five and continued for an hour. I then managed to get back to sleep until eight, when I headed for a shower in the fabulous bath. It usually takes me ages to determine how the fittings work, and exactly where the dial must be to produce a pleasantly hot mix, but the old-fashioned taps made this an exception. In no time I was quietly singing to myself as I poured a generous helping from the expensive-looking liquid shower gel bottle left in the soap dish, and worked up a rich, smooth, lather from head to toe, while soaking up the heat from the powerful water jets.

A perfect shower is a thing of great pleasure, so it was some time before I reluctantly turned off the taps and took the massive towel from the conveniently placed rail fixed to the sidewall near the end of the bath. I was snug and warm, in a safe environment, happy, relaxed and wrapped in a fluffy white cocoon. What could possibly go wrong?

Since the towel rail was near the end of the bath, that was where I now stood. Coincidentally, the shower curtain only ran the length of the bath, so the end was open, and the obvious exit route. Without thinking, I launched my left foot toward the back of the bath, having to adjust midway to clear the long, sloping end. Probably just after my foot passed the point of no return, I remembered the large step up into the bath, and realised it meant a large step down when getting out.

In a desperate attempt at aborting my exit, I grabbed for the towel rail, gaining a good, firm grip and a nano-second's relief before I realised it had snapped out of its fittings. I twisted in mid air, to clutch at the shower curtain, but it was slippery, and I had zilch success. While staring at the material sliding through my fingers, I was horrified to see an image of the devil staring back at me, pitchfork in hand. Instantly, Phil's words came crashing into my mind: 'You're going down.'

Having twisted left for the towel rail, then right and back for the curtain, somehow I now managed to twist left and forward again, to see the sink approaching my chin at an impressive rate, yet in slow motion. Three things now happened together.

My left foot hit the floor and started to slide back towards the bath, while my right foot lifted from the bath and began to follow the rest of my body over the end. I felt my body give a massive jerk away from the sink, using my stomach, neck, and any other muscles in the vicinity.

All this resulted in me watching the ceiling as the sink flashed past, presumably within an inch of the back of my head. As my left foot slid away, my knee, hip, bottom and back took turns in breaking my fall, and I found myself lying on my back on the floor, left leg resting against one of the bath supports and right drifting down to join it. My head was under the sink, right hand empty, while my left still clutched the useless, and somewhat bent, towel rail.

I lay in stunned, shocked silence, attempting to piece together what had happened and summoning up the courage to inspect my injuries. My daughter, Claire, constantly warns me of the dangers of moving someone with a spinal injury, and that people in shock often feel no pain, so I wasn't going to do anything hasty. I tried wriggling my toes and flexing my fingers, and all were fine. I gingerly moved my hands, arms, feet and legs: fine. I pulled my knees up towards my body, sliding my feet along the floor: fine. Expecting to die or pass out in agony, I slowly turned my head from side to side, experimented with flexing my back, and very slowly stood up: absolutely fine! I stood and looked in the mirror: dreadful sight, but nothing I wasn't used to – not a single sign of what I had just been through – not a scratch, scrape, red mark, bruise… nothing. Amazing… *Amazing*! Phil! *Dephil*!

I recalled the devil on the shower curtain. Who would have such a curtain round their bath? A closer look was called for, and revealed all. The curtain was covered in images of the sea – fish, shells, mermaids, octopuses, and just one, of… Neptune, trident in hand. Naturally enough, mixed with relief, this set me off chuckling to myself and, while I replaced the bent towel rail, a great new song for the walk sprang to mind:

I get knocked down
But I get up again
Ain't never gonna keep me down.

Returning to my room, I noticed for the first time the cheerful-looking stuffed fox grinning at me from the sideboard. That was the final proof: Lucky Fox was going to win. Dee / Phil would be Dee-feeted, blisters and all. I was still singing when I walked into the breakfast room.

'What're you so happy about? Didn't you see that stupid looking stuffed fox in your room – looks like its luck sure ran out. And what was all that crashing and bumping in the bathroom?'

'I just had a near-death experience, with my life flashing before my eyes – well, the sink anyway – due to slipping as I stepped out of the bath. I think I must have done a somersault with two twists and pike, landed with hardly a splash – gave myself 9.7.'

'And you thought Tubthumping summed it up?'

'What?'

'That was what you were singing, wasn't it? "I get knocked down, but I get up again. Ain't never gonna keep me down".

That's Tubthumping by Chumbawumba.'

'Who?'

'Chumbawumba.'

'Come off it, Phil. I've never heard of them, and tubthumping hardly fits the words. You're making it up.'

'Would I lie to you?'

'Seriously? That's spooky. I just thought the words suited the way that despite how hard you tr… Is that my cup? Pass the pot, will you?'

'Eh? Oh, yeah. Now, as you were saying?'

'Er… well… just that… that despite how hard you try… something always goes wrong… and you get knocked down, but we always manage to bounce right back up. Like in that song. But for its title to be Tubthumping when I just came so close to being given a right good thumping by the bath tub is just such an incredible coincidence. Makes me think there must be more to it – perhaps its a sign.'

The hairs on my neck were standing up and shivers racing up and down my spine. I had managed to cover up nearly telling him I thought he was a close bosom buddy of Beelzebub, but this was getting silly.

'You could be right, it might well be a sign – that you are a clumsy beggar who can't even get out of a bath without wrecking the place. I'm pleased you're okay – I thought someone was attacking you and I was going to have to eat your breakfast.'

Thanks, Phil.

Mind you, breakfast was superb, and I wouldn't have blamed Phil for trying to eat two, not that he could possibly have got through them. Top quality furnishings and crockery were matched by an equally high standard of cooking. There were ten different cereal choices, including homemade muesli, followed by a full traditional English fry-up, toast and marmalade, and a bottomless pot of tea. Ideal for walkers, since it gives you that warm, smug tummy feeling well into the afternoon.

Mrs B&B also had great news – she had fixed us accommodation that night with one of her friends, and her husband had agreed to take us and our bags there straight after breakfast in his car. We exchanged glances, which basically agreed that she had indeed fixed us, and we would say nothing about our other booking for now, then discuss our options while packing. Not surprisingly, we decided to take the cowards' way out, going along with Mrs B&B's arrangement, and sorting out the Youth Hostel later – of course we wouldn't just not turn up.

Mr B&B parked his car outside the front door, and we were somewhat taken aback by the amount of rust and number of holes in the bodywork. When I lifted the tailgate to place my rucksack in the boot, and the support arm fell off, completely rusted through, we were in two minds whether to risk the trip. Perhaps we had misunderstood when Mrs B&B said her husband had agreed to take us in his car – was it meant to be a warning?

The engine rattled and screeched awfully as we pulled away, and at every turn or hill, while the brakes seemed extremely spongy so I lived in fear that we wouldn't slow in time to take the next curve, instead going somersaulting down into the harbour. Somehow we reached our new accommodation, Boak House, which proved to be very close to the youth hostel, so while Phil checked into our rooms and dumped our gear, I went to do the decent thing and cancel our booking. I expected a rollocking for mucking them about, but actually received a very friendly 'no worries' greeting, and list of phone numbers of all Youth Hostels on our route.

Back at Boak House, I asked if there was anywhere to dry our tent, and was shown the sun lounge, which was a five-foot wide lean-to extension running along the front of the house, filled with washing lines, and absolutely ideal. The outer wind-sheet was not particularly wet, so I hung it straight over a couple of lines, but the main tent was sopping, and I was concerned that it would not dry thoroughly. I had noticed that there was quite a stiff wind coming round the corner of the house, so stood for ten minutes in the front garden, with the wind blowing in the door of the tent, filling it like a hot air balloon. The sides billowed and flapped crazily, with me tightly gripping the fabric and attempting to steer away from bushes and walls, being almost pulled off my feet on several occasions, and having immense fun.

Suddenly we realised it was 12.00p.m., our bus was due in twenty minutes, and we had done it again, despite our solemn vow only the night before to properly plan and impeccably implement. We grabbed our gear and jogged off down the hill and along the sea wall, confident we would be okay since we could see the bus waiting at the stop. Our confidence took a severe dent, however, when it left while we were still a couple of hundred yards away. We arrived at the bus shelter a few minutes before the scheduled time, looked on the timetable there, and read 12.10p.m., not 12.20p.m., with the next one not due until 2.10p.m. We turned to each other and chorused 'you must have misread the timetable.'

We seemed to have little choice but to walk Coverack–Helford instead of Helford–Coverack, which would mean the tide would be wrong at Gillian Creek crossing. After fifty yards, and just before we lost sight of the bus shelter, another bus arrived: the 12.20p.m., which, for some reason, was not included on the timetable in the bus shelter. Lucky Fox was back on form.

We were used to highly skilled drivers belting along the winding country lanes at break-neck speed, so it was an interesting experience to find ourselves with a learner driver being talked round the route by an instructor. He drove very cautiously and took each bend as though there could be something coming round it from the opposite direction. It made us appreciate the skill of the other drivers, but also ponder whether they were actually safe.

At Helford, we were dropped some three hundred yards inland of the ferry crossing point and our route led further inland, but we decided to walk to the ferry for a look at where we would be heading the next day. A visit to the post office and general look around Helford ensured we managed to waste an hour before finally setting off up a very muddy footpath along a wooded stream towards Manaccan. We were taking the inland route to avoid wading Gillian Creek, which we had spotted as an option when panicking after missing the first bus that morning.

We managed to misread the map, and left the footpath too early, ending up near a farmhouse with two loose Doberman Pinschers, which came rushing towards us barking loudly, menacingly baring their teeth. We had our walking sticks to protect ourselves with so, being intrepid adventurers, were not in the least put off by them, honest. It was pure coincidence that we now decided to follow a track that gave the farm a wide berth, and led to a winding road. We finally arrived in Manaccan at 2.30p.m., nearly two hours after getting off the bus, having advanced less than a mile towards our destination.

We found the footpath leading down to Gillian Creek and the crossing into Carne, and met a woman walking two dogs. Phil stroked the dogs, and I asked if I could stroke the owner. She replied 'that all depends', gave me a great smile, and walked on. By the time I had realised what she had said, she was already fifty yards away. I turned to Phil:

'Did you hear that?'

'Yep. Good old 'mouth and no trousers' strikes again.'

Several steep ups and downs led into Porthallow, where we found the Five Pilchards pub and a beach café, both closed between 2.30p.m. and 6.00p.m. Not normally best news at 4.00p.m., but it was in this case because it forced us to look elsewhere and we found the excellent Taranki Tea Gardens opposite the Post Office. Beautiful flower-filled gardens, including a glass lean-to and pagoda style patio area, were home to nine cats, each with nine lives. Phil managed to find six, learn their names, ages, likes and dislikes, physical disabilities and distinguishing features, and repeat them to me every time one appeared while we ate our lunch.

The waitress recommended crab as the speciality of the house, so we each partook, sandwiches for Phil and a massive ploughman's for me. Phil ordered homemade (by Dorothy, the owner) fudge cake for dessert, which the waitress brought with the greatest apology I have ever heard: 'I'm sorry it's so large – it's the last piece and I didn't want to cut it.'

Phil let me have a measly quarter – 'think of your diet' – and it was absolutely delicious. Rich and moist, with a melt in mouth fudge sauce and thick, dark chocolate top, accompanied by a mound of Cornish cream. While we savoured the cake, the waitress informed us that the flowers in the garden were only part of a vast range of exotic plants grown by Dorothy, and that she had won several prizes for them. Before leaving, we bought some Dorothy-made scones and rich fruitcake for future consumption.

The path was forced inland due to the large number of quarries, both disused and current, between Porthallow and Porthoustock. We passed a farm with two fields, each containing a large flock of ostriches, and followed the road down to the beach, finding ourselves facing an aggressive sign claiming it to be private. Two teams of scuba divers were making final preparations for a dive on a local wreck, so we watched them wade into the water and disappear, then followed an ancient rail track up onto the headland, through a disused quarry, and on to Porthoustock, which we agreed to be the worst place we had seen on the whole walk. Awful, massive, concrete slabs were scattered along the sea front, there was rubbish all around the stream, and a sign asking for donations towards beach maintenance and improvement, which was sorely needed, but fell on deaf ears.

We pressed on, up a very steep road and across some fields into a working (but not then) quarry area with massive machinery, conveyors, chutes, and piles of stone. Signs pointed out the route, but it felt extremely dangerous, bleak and hard, and we were very pleased to round the headland and see Coverack only a couple of miles away.

It was raining very heavily when we set out for our evening meal, down the incredibly steep road from Boak House to the harbour area. The road markings seemed to have been recently painted, and beads of water sat on their shiny surfaces, so I kept well away. Phil must have been in a particular hurry to get to the bottom of the hill, however, because he decided to cut the corner at the bend halfway down, stepped on the glistening white line, and attempted a re-enactment of my bath exit that morning. Happily, he escaped without injury, but even more happily, not before sliding about thirty yards down the road on his back and overtaking a group of lads from the youth hostel, a couple of whom looked as though they thought he was doing it deliberately, and might give it a try themselves.

We decided to eat in the Fodder Barn, since it was close, open and had space, the only drawback being the interior lived up to the name. There was a good menu, from which we both chose French onion soup, Phil followed with pizza, prawn creole for me, while we shared two bottles of wine. I know it is a cliché to say that the staff looked very young, but the only two we saw cannot have been older than ten, yet carried out all front of house activities.

It continued to rain heavily all evening, with frequent top quality lightning. Throughout the meal, Phil kept making trips to the phone box, trying to book us in at Pendennis Castle Youth Hostel for the next night. Although gone for ages, he always returned with a long face, stating that he hadn't wanted to let his meal get cold or allow me too much freedom with the wine, so had given up but would try again later. He explained that there was always a queue and everyone seemed to have plenty to say once they finally got their turn. On two occasions he had managed to get to the phone, but the number was engaged. On another, the person in front finished their call and hung up, but before Phil could get his money out the phone started ringing and a local lad appeared out of a shop doorway saying 'That will be for me.'

Given how heavily it was raining, I was perfectly happy with the arrangement, but felt it my duty to at least offer to take a turn. Phil was keen to keep going himself, however, presumably feeling it was time he did something towards the holiday.

On the way back to Boak House, I noticed the phone was free and got through to Pendennis Castle first time, with Phil looking rather sheepish. Then followed a very entertaining stagger up the hill, hindered as we were by the pouring rain, slippery surface and bottle of wine inside each of us.

I woke to the usual gull chorus of squawks and squabbles, but this one was slightly different. One of the gulls seemed to have whooping cough and couldn't finish one squawk before the next started. It seemed to be getting slower and weaker, and I became as fascinated as I had been on the steam train trip when The King was struggling up Wellington Bank. Eventually, it stopped half way through a pathetic squawk. I resisted the temptation to look outside for a dead body but then couldn't get back to sleep through wondering whether it was okay.

Being in the attic, with the sun streaming in through the

window, my room soon became unbearably hot. I gave up any attempt to get back to sleep and went out to the sun lounge to re-pack the tent, which was now perfectly dry, while taking in the fantastic views over beach and harbour.

As we made easy work of yet another quality full English breakfast, we chatted to a very tall, athletic German couple sitting at our table. They were also on 'the path', so we swapped tales and information, until Phil took over by giving us a free morning lecture: 'Why no English want to be in the Euro', including an aside on 'how Germany is going down'. They took it well, mainly agreeing with him (perhaps they were Austrian, or Dutch), but even I couldn't stand it when the subject changed to 'what is wrong with the German football team', so decided to drag him off.

Being late for the bus, we had to pack and pay in three minutes, then stride the mile to the stop, in spitting rain but with no time to put our jackets on. Our bus was not listed on the shelter timetable, but we were seasoned travellers, so didn't worry, our faith being rewarded when it duly arrived. The driver zoomed off as if making up for the previous morning's learner, taking bends on two wheels, and braking and accelerating madly. Soon after St Keverne, he was buzzed on the radio and told he had missed two passengers who had been waiting in the shelter – 'how am I supposed to see them in a shelter?' – and must go back for them. We were not amused, and nor was the driver who muttered to himself 'I'll be late now' more as a threat than complaint, and so it proved. He crawled along, taking every bend slower than the last, changing gear slowly and methodically, then barely touching the accelerator, and letting cars and slow farm vehicles out of side turnings.

Back at St Keverne, the missed passengers, two young girls, just stepped on board and asked for their fares as if nothing had happened, neither thanking him for coming back, nor moaning about being missed in the first place. Having made his silent protest, the driver reverted to character, and we were soon bouncing around in the back with broad grins on our faces.

Our plan was to take this bus into Helston, then another to Falmouth, dump our gear at Pendennis Castle, catch a third bus back to Helford, then walk back to Falmouth. I persuaded Phil this was too much messing around, and we should cut out the round trip to Falmouth. Instead we would bus straight to Helford, then walk to Pendennis Castle carrying our gear: it's only nine miles. At Helford we headed for the ferry point, where a timetable informed us we had an hour to wait, but then we spied some electronics, a plastic board with small holes, and a sign, which instructed:

> PRESS RED BUTTON FOR SEVERAL SECONDS (buzzer)
> THEN PRESS GREEN BUTTON WHILE SPEAKING INTO MICROPHONE HOLES.
> RELEASE GREEN BUTTON AND LISTEN FOR FERRYMAN'S REPLY.

A family of four had been trying this with no effect, but we still insisted on having our turn. It felt odd to press various buttons, talk into a row of small holes, and receive no response whatsoever, particularly knowing that several other people had just done exactly the same, with the same result. Were the buttons working? The microphone? Was the ferryman on duty? Awake? Likely to respond? Was he shouting into a row of holes on his side?

With so many potential missing links, we looked for an alternative approach, and soon found another sign, on a large white semicircle, which read:

> OPEN THIS SIDE TO REVEAL ORANGE CIRCLE TO SIGNAL FERRYMAN.

We did this. Nothing. We started to play around, sending supposed Morse code messages by opening and closing the shutter. Whatever we did, we received no response, but at least we had kept ourselves amused for half an hour, and only had another half hour to wait until the scheduled ferry time. By now, the other four had already left, having decided to drive round. They were only going across to hire a yacht for the afternoon – they were not walkers, or anything important.

The ferry arrived exactly to schedule, the ferryman taking a quick look at the plastic communication board and saying 'Oh, is that out of order again' with a glint in his eye which hinted at another explanation for the lack of response. It was a beautiful smooth crossing to Passage Cove, with no problems apart from me almost falling in (tip: don't wear your pack while getting on or off a boat. It makes it much more likely you will fall in, and if you do, much less likely you will get out again).

Passage Cove proved to be a small jetty / boat hiring place, with a couple of cafes and tourist shops. Phil fancied a coffee before we started, so we sat outside the yacht club watching the world go by, which soon included the family who had decided to drive round. They were refused a yacht because it was too windy, which they seemed to find rather odd as wind is a basic requirement of sailing, so decided to give up and play golf instead.

After a second coffee, we set off. Judging by the map, the path was obvious and easy: stay next to the sea all round to Rosemullion Head and beyond. Fairly soon, we were on a road, in heavy woodland, near Greeb Rock, looking for a left hand bend at which the footpath continued straight on. We came to a path which led down towards the sea, was on the right side of the road, and on a bend, the only problem being that a South West Coastal Path signpost (brown with a yellow acorn logo) seemed to be pointing up the road. We decided the sign must have been angled badly, and that the correct route must be down the path.

A steep four hundred yard descent brought us out onto a small rocky beach with clearly no exit at the other end. Nothing for it but to turn round and pound back up the steep hill. Almost immediately, however, we noticed a scramble path leading from the one we were on very steeply up towards the road. I suggested we go for it: it would save three hundred yards each way, and there seemed to be plenty of trees etc. to hold on to.

As we scrambled up, the path worsened. Recent heavy rain made it very slippery, and it became even steeper, so that I was soon climbing, using the undergrowth, tree trunks and branches, and Phil's head and shoulders, to stop me sliding straight back down. Two songs came to mind: Paul Simon's *Slip Sliding Away – the nearer your destination, the more you're slip sliding away*, and an old Country and Western one *One Step Forward Two Steps Back*.

Ultimately, all sign of previous human endeavour disappeared and, feeling like a west country Indiana Jones I found myself at the bottom of a twelve foot cliff face made of loose, slippery rock, the top of which was overhanging. Even a real-life Indiana Jones would have balked at it, and following Stackhouse Cliffs I was in no mood to give it a try.

Phil was not amused when I relayed this to him, and as we slipped and slid our way back down to the beach path, constantly expressed his gratitude and continued faith in my leadership and orienteering skills.

The existence of the steep scramble path to nowhere brought a vision of group after group of walkers deciding the signpost must be wrong, marching down to the beach, starting back up, spying the scramble path, forcing their way up, then sliding back down. Someone must have made the first attempt to force a quick route back to the path, failed and turned back, since when increasing numbers had seen the steep path up, and done the dead-end loop, thereby keeping the path fresh and inviting. I pointed this out to Phil:

'Everyone must make that mistake.'

'Oh, that's all right then.'

Once back at the junction between this path and the road, we marvelled at the clarity of the sign so obviously pointing up the road, and our decision to ignore it. Sure enough, five hundred yards further up the road we came to the right corner and path, again clearly marked.

En route to Pendennis Castle we were caught by a vicious little rain cloud, so we stopped for a cup of tea and lunch fit for a gorilla – apples, carrots, coconut, scones and the rich fruitcake from Taranki Tea Garden.

It was 4.30p.m. when we reached the top of the sharp hill leading to the castle gate, to discover that the youth hostel was closed until five. We were, however, allowed to enter the grounds so dumped our gear and went sightseeing. Pendennis Castle was built by Henry VIII to protect the Carrick Roads and Falmouth Bay area, with a smaller castle, St Mawes, on the opposite bank. Between them they could cover the full width of nearly a mile, cannon in those days only being accurate for about eight hundred yards.

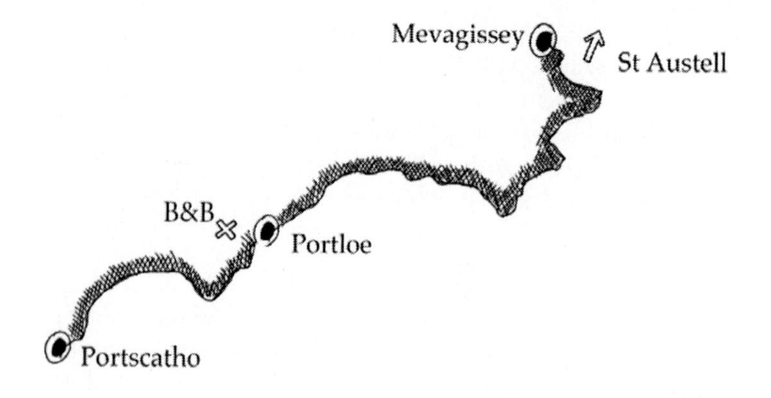
Mevagissey
St Austell
B&B
Portloe
Portscatho

Pendennis Castle has been developed and maintained ever since, including being used as a strategic base in both World Wars. Armaments on display included cannon and artillery throughout the whole period, with several mock battle scenes accompanied by dramatic sound effects of gunfire, shouts and screams, flashes, fire and smoke, and the acrid smell of gunpowder. Views from the battlements were fantastic all round, including scores of ships and smaller craft taking full advantage of the late afternoon August sun.

We settled into our six-bunk dormitory, freshened up, Phil proudly putting on his 'glow in the dark' eclipse T-shirt, bought for five pounds the day after the eclipse. On the front was a large fluorescent circle representing the sun's corona round the eclipsing moon, and on the back, a full listing of the sites and dates of all twentieth century eclipses, headed 'The C20 Eclipse Tour', again all in fluorescent paint. Our evening meal was soup, Chile con carne and apple crumble with custard or ice cream, all great, traditional stuff, very satisfying and pleasingly reminiscent of the best of school cooking. We lapped it up, while agreeing to walk into Falmouth to check the location and times of ferries to St Mawes, thence a place called Place, the start of the next day's walk. We could have a general 'sticky beak' look round, and drink a couple of pints in yet another of Phil's favourite pubs.

It was one and a half miles into town, through a mixture of modern yacht marinas and accompanying hotels / timeshares, modern and old docks, and the TA centre which looked like a prison, including razor wire on all the walls. Falmouth town centre had a shopping area consisting of very old buildings and narrow roads which, although it was probably quite empty, felt very crowded to us as we actually had to avoid bumping into people, and wait to be served in the pub.

For the last couple of days, we had been toying with when and where to end our walk. Theoretically, we had until Saturday evening, but Phil was now keen to finish at Mevagissey, which was only two day's walk away, giving a Friday finish, and an obvious Thursday night stop at Portloe. While we finalised our plans and reminisced, two pints became three, then four, so we found ourselves running back up the hill to beat the locking of the main gate at 11.00p.m. I was rapidly getting over my cold, and Phil was catching it, looking increasingly shattered all the time, so he was the one puffing at the top while I made encouraging remarks.

We had been told by the gate keeper that it was still possible to get into the hostel after eleven, provided you could climb over the castle walls, which was easy at a certain point. He hadn't said where that certain point was, however, and we didn't fancy spending half the night testing likely looking spots, still not quite over the Stackhouse Cliffs experience. I noticed a white camper van with a personalised number plate: DEE 6 L, but thought better of pointing it out to Phil.

In dormitory four, all was peace and quiet, with the other inmates, working on walker's time, unlike us, having turned in some time ago. Although Phil doesn't know the meaning of 'whisper' I think we made a pretty good job of grabbing our gear, doing the usual, undressing and getting into bed. Each bunk had a personal light, but we only used Phil's, so as to minimise disturbance.

My bunk was above a guy who looked at least eighty, which made me think of myself as a legitimate youth hosteller. It was impossible to climb in without rocking both bunks, and I noticed he was looking at me, with a pained expression. I was not sure whether he was annoyed at being woken or just scared I would come crashing through the impossibly thin wooden slats any second, so gave a friendly grin which he responded to by letting out a sigh and turning to face the wall.

As his final act before switching off the light, Phil took off his sweater, the room was filled with an eerie glow from his eclipse T-shirt, and another sigh was heard from the old youth below me. I just managed to fight a fit of giggles, which I'm sure Phil would have joined in on, fuelling mine, had he not fortunately turned to face the wall as he laid down, remaining blissfully unaware of the light he was bathing us in.

Phil was not in his bunk when I woke, and didn't appear until after the breakfast starter, looking a right mess, still unwashed and unshaven, and bleary-eyed. He explained that he had felt a bit rough during the night, and gone to sleep in the lounge so as not to disturb anyone. I told him not to worry, I had eaten his porridge for him, and it had been just how Goldilocks would have wanted it. When we had finished, I noticed an untouched plate and asked if I could nick a sausage. The cook stopped me, saying that someone had booked a late breakfast.

Down the hill into Falmouth for the 10.15a.m. ferry, which arrived bang on time. Having let those on board off, we went to step on when the captain said:

'We are just going to refuel. Back in seven minutes. Honest.'

That was too close to my use of the word for comfort, so I thought I best check:

'Are you refuelling yourselves or the boat?'

'The boat. Just getting some ocean lotion.'

They returned as promised, and we climbed on board, remembering to take our packs off – at last we had actually learnt something. Soon after departing, with our ticket and money safely in his pocket, the captain gave us the news:

'You two look like you want the Place ferry after this,' we nodded, 'well, it's not running. Engine trouble. Probably be okay tomorrow.'

We studied the map. It would have been easy enough to walk, carrying our gear, from St Mawes to Portloe, but much of the route was inland, looked relatively boring, and was not on the official Path, which we felt would be an anti-climax for our penultimate day. We could have taken a taxi from St Mawes to Place, then walked with all our gear to Portloe, but Phil was somewhat reluctant to do too much carrying, given his terrible cold. Ultimately, we decided to do one of our complicated manoeuvres: taxi from St Mawes to Portloe, find a B&B, then either taxi back to Place and walk to Portloe, or vice versa.

We managed to book a taxi in St Mawes, but with a half-hour wait. This gave us a chance to have a quick look around, which we took the lazy way: let our eyes do the walking while we sat having tea and a bun in a harbour-side teashop. I noticed a pub with the name 'Idle Rocks' and pointed it out to Phil as reminiscent of his sex life, which resulted in him giving me a very strange look, shaking his head, and starting a conversation with the shop assistant. Okay, it had been a bit cruel, but how could I resist? As we were getting in the taxi, I suddenly realised I had left the map in the phone box, which Phil pointed out as reminiscent of how we became lost on our first morning. Touché!

We now enjoyed a rare event: all our taxi drivers had been very friendly and helpful, but this was the first who was pleased with his lot in life and had a happy and successful marriage. He and his wife worked as a team, with her taking the bookings, and both appreciating how hard the other worked. We saw their teamwork in action when he phoned his wife to get her to ring around Portloe B&Bs to find us a bed for the night so he could take us straight to it.

A great idea, which deserved better luck: none of their usual contacts had any vacancies. Not easily beaten, he drove us around Portloe until he found one, at Spear Point, which, as usual, was excellent. It had two life-size statues of Laurel and Hardy dressed as tramps and seated at a park bench – the similarity between us and them was obvious. Laurel's boots had holes in, and I fancied I could see a blister oozing through.

The decision whether to walk or taxi back to Place was made for us when the taxi driver said he had other bookings for the rest of the morning, but was fairly free early evening. He suggested that we phone his wife from the pub at Portscatho, two-thirds of the way to Place, to arrange a time for him to pick us up. Even I found it hard to begrudge the fare.

We treated ourselves to a free coffee in our rooms before setting off on our walk round Roseland, the first two miles of which were terrible, back to the up, down, up, up, down of a few days ago. Shortly before Nare Head, however, we hit a patch of rolling grassland, which relieved our aching leg muscles. From Nare Head, the view was not only spectacular, but also allowed us to indulge in one of our favourite activities: spying out most of the day's walk, and matching it to the map. Across Gerrans Bay with its nice mix of long sandy beaches and high craggy cliffs, we could see round to Portscatho and on to Greeb Point and Killigerran Head. The map showed this to be hiding St Anthony Head, which lies at the end of Carrick Roads, opposite Pendennis Castle, our previous night's accommodation. From there, we would follow the path round the cliffs for the final mile into Place.

'Here, John, did you know Anthony Head is the Gold Blend man?'

'Fascinating, Phil.'

'And he's now in Buffy the Vampire Slayer.'

Lucky man. '*Amazing*.'

The walk down from Nare Head was fabulous and carefree, on springy grass in glorious sunshine, but as we approached Carne Beach we could see it was crowded and likely to ruin the spell. Then we spied our potential salvation, or downfall: an obviously five star hotel set in its own private lawns at the back of the beach: Nare Hotel. Like a pair of complete innocents, we ignored the certain refusal, given our muddy boots and dirty, sweaty clothes and faces, opened the gate and pressed on.

Now, we could see the top quality decor, beautiful furnishings and food, genteel, well-dressed clientele, and black tie waiters hovering around, and we both knew there was absolutely no chance of them allowing us to disturb the ambience, leaving nasty stains on their furnishings and reputation. We had passed the point of no return, however, so rather than slink away with our tails between our legs – possibly literally in Phil's case – we marched right on, past the oh-so-tempting swimming pool and on to the terrace, heads held high, as though we were doing them a favour boosting their image by our mere presence. A waiter came to meet us as we approached the tables, and I tried not to flinch.

'Hello. Would you like to see the menu?'

There were some price traps for the unwary (cream tea six pounds fifty) but we selected an old favourite: Moules Marinière!

Our thoughts were increasingly turning to how our trip was drawing to a close, we tended to use the past tense when discussing it, and agreed that, while we had thoroughly enjoyed it, we seemed to have been away for ages and it would be good to get back to reality. This taste of luxury, coupled with the reminder of our Penzance days, heightened such feelings, and we began to reminisce. Phil agreed that it was 'amazing' Newquay had only been two and a half weeks ago, but went quiet when I said I couldn't remember what Dee looked like – possibly brought back some memories he preferred to forget.

It was so good to have some real luxury and pampering for a change, the waiters seeming to enjoy having us there, and we lingered for an hour and a half. All good things must end, so eventually we climbed back into our packs and reluctantly shuffled off.

Cracking along at a great pace, which was just as well given our lost time, as we neared Portscatho, it began to rain very heavily. One of my tendencies in such situations is, in the hope that it will only be a shower, to delay putting my waterproof trousers on until my legs are already soaking and it is pointless to do so. Consequently, on bursting through the door of 'The Plume of Feathers' and taking off my jacket, a three-inch section of my shorts were saturated, the rest being dry. This attractive combination caught several eyes in the bar, and I expect it will become quite fashionable, possibly rivalling wet T-shirt competitions as a source of entertainment. Phil did his part by making the general announcement that it isn't easy walking with a catheter, leaving me no choice but to join in: 'We normally walk incontinent, but decided to stay in England this year.'

Having found ourselves in a pub in a storm, what could we do?

By the time we got round to ringing the taxi, he was fully booked for the evening so could only pick us up within the next ninety minutes. Once again, our casual attitude towards planning found us pounding along, this time down to and round St. Anthony Head, with far too little time to do justice to the fantastic views across Carrick Roads to Pendennis Castle.

That evening, walking down the hill from Spear Point into Portloe, we came across a pub, purporting to sell great evening meals, which proved to be the case. We ordered giant prawns for starter, and cod and chips main meal, only to discover the cod was off: 'I don't want it then' did not raise a laugh. Phil managed to keep to his fishiness with fish pie, while I branched out with Pork and Apple Hotpot.

The starters were enormous, with four giant, giant prawns, vast quantities of salad, and a hot roll. Phil whacked his back in no time, but I struggled to maintain my dignity while levering the shells off. All the time, we were having gentle banter with the barmaid, Billie, whose boyfriend had recently left on a world tour, leaving her to finish her degree. Phil did a disappearing trick with his fish pie, almost including the dish, while I worked my way systematically through my hot hotpot.

Since this was to be the last night of the trip, we decided to celebrate with several pints, so Billie eventually had to kick us out well after closing time. Despite the earlier rain, the sky was now completely clear, and stars shone brightly down. Noticing my interest, Billie said she lived in Truro, so would be driving to the top of the hill if we fancied a ride, the only problem being her back seat had a load of gear on it, so there was only room for one. Phil said he wanted to look at the boats in the harbour, but I decided it was an opportunity not to be missed.

On the drive up, being my usual nosy self, I turned to see what gear was filling the back seat – just a coat and large, soft blanket. Billie laughed, squeezed my thigh and said 'Two's company, and we don't want to get cold up there, do we?' Yes, definitely an opportunity not to be missed.

Friday, 20th August. Our last day!

Following a fabulous shower, we had what had become an expectation, a top quality full English breakfast, but this time enjoying fantastic views across the harbour, including a reflection of the sun shining in a deep blue sky dotted with wispy cloud. Everything seemed perfect, as though The Path was tempting us to stay, or at least ensure we regretted leaving and came back soon.

Despite our having thirteen miles to do, and having the cold full on, Phil had agreed that backpacking was a must, so we were able to start walking without worrying about taxi or bus. Other walkers had warned that the first two miles were ridiculously hard, suggesting a much flatter inland alternative. Okay, we were wimps, but the thought of killing ourselves in the first two miles, with eleven still to go, did not appeal. We set off inland.

If that was the easy route… the first half-mile was straight up a very steep hill – initially on the same road Billie had taken the previous night. Sure enough, her car appeared round one of the bends, and I knew I was in deep trouble – where was that fox when I needed it? I tried to brazen it out by waving my stick and calling 'Hi', but she sped straight passed, missing me by inches, screeching to a halt next to Phil while opening the window. I wanted to zoom off up the hill, but couldn't stop myself listening:

'Hope you had better luck with the boats in the harbour – at least they go up and down when the tide's in.'

Phil clearly didn't understand what she was getting at because he just stood looking at her with that dumb open-mouthed expression he is famous for. I winced.

'I don't know which of us was taking the other for a ride.'

More blank looks and winces.

'We both jumped out of the car, I got into the back and turned round to find him wandering off across the field, stargazing. First bloke I ever knew...

At last Phil cottoned on, and immediately burst out laughing.

'Don't tell me he bottled it again.'

'Bottled it? I don't know whether he bottled it or throttled it, but he sure didn't share it with me.'

Phil collapsed on the car bonnet, howling with laughter, trying to speak but unable to get out more than a squeak, like Brian Johnson in the famous cricket commentary. Eventually, Billie caught a few words:

'Mouth and no trousers. What do you mean mouth and no trousers?'

More giggles and squeaks.

'What is he? A monk or something? Are you two an item? He was chatting me up all night – how was I supposed to know he really wanted to look at the stars.'

By now, Billie's voice had softened, and I had walked back down the hill to join them:

'When you two have finished, perhaps we could get on with this walk?'

Phil made an innocent enquiry:

'So he didn't show you his Big Dipper or any shooting... stars?'

'Not even his Ursa Minor, and we certainly didn't go the

whole Milky Way, either.'

We were all smiling now:

'Anyway, how was I supposed to know when you spent five minutes locking the pub door while the locals disappeared, made a feeble excuse to get rid of Phil, and offered me a lift to share a heavenly experience, that you were after my body?'

'I'll say this for you, John, you sure are different.'

'Any chance of giving me a ride – to the top of the hill?'

'Now that's hardly the spirit of the Path, is it?'

'I guess not.'

'Well, it was nice nearly knowing you. I've got a bar to clear and glasses to wash, while you two carry on your fun and games.'

As Billie disappeared down the hill, I heard Phil mumble something like:

'Just wait 'til Dee hears about this.'

'What did you say?'

'Shi...nothing.'

'Yes you did. I distinctly heard you mention Dee.'

'Er... I... em...'

'Telling her about this.'

'I was just thinking...'

You are still in contact, aren't you?'

'Well, er... you know... sort of... erm...'

Suddenly, all was revealed:

'You have been seeing her at night! You crafty beggar. That's where you've been sneaking off to every night, isn't it?'

'Erm... well... '

'Come on, admit it. She didn't just want a one-night stand – you have been standing every night, haven't you?'

'Alright, I admit it. Not every night, but whenever I could get away.'

'You have been seeing her! Ha ha, that's fantastic! Well done, my son. Here's me feeling sorry for you being given the old heave-ho after a one-nighter, and you have been burning your candle at both ends, and a bit on the side. No wonder you have been looking more knackered every day.'

Phil looked relieved to have confessed, and broke into a

massive grin.

'Okay, it's a fair cop, I confess all. I have been dipping my wick at every opportunity.'

'I didn't suspect a thing – you must have been brilliant.'

'You must have been blind, more like.'

'What do you mean?'

'Oh, for heaven's sake. What did you think I had been doing every night?'

I could hardly say 'Satanic rituals and sacrifices' so changed tack:

'So you sneaked off each of the last three nights in Penzance.'

(with some pride) 'Yep.'

'While I was getting my beauty sleep, you were getting your…'

'Yep.'

'So that's why it took you four hours to have a pee that night. I thought you had nodded off on the throne or something. No wonder you gave me that odd look when I said that 'Idle Rocks' pub name reminded me of your sex life. Here's me living like a monk for three weeks, while your rocks have taken a worse pounding than anything Geevor Mine would offer. And I'm the one with BR! But why didn't you tell me?'

'I didn't want you to think I was going to let you down again.'

'Oh, for heaven's sake. The blisters weren't your fault, I've had a great time, and I don't feel let down. Suddenly I recalled our conversation before I had returned to London for the party.

'You aren't still fretting about Kilimanjaro are you?'

'To some extent, yes. I wanted this walk to be a kind of "thank you" for at least trying to help when I really needed it and no other beggar lifted a finger or even said an encouraging word. It would have been a great "thank you" to have cleared off with Dee, leaving you alone in the middle of darkest Cornwall.'

'We could have come to an arrangement, you could have spent the evenings and nights with her, and the days walking with me – if you weren't too shagged out.'

'That's what I have been doing.'

'Fair point, well made, but why the secrecy?'

'I don't know. I guess it was part of the fun, seeing how long it took you to stumble across what was happening. You must admit, it is incredible that you didn't suspect a thing when we have been sharing a tent or the same B&B for a week, and I have been disappearing all night, every night.'

'Including Coverack?'

'Both nights. First, I nipped back out when you had gone to bed, and the second used all those phone calls as cover. Dee was in the Youth Hostel both nights, but agreed to let me actually sleep the second one after a quickie behind the harbour wall.'

'You crafty little sod. I thought it was odd that I got through first time. What about Pendennis Castle?'

'Who did you think had booked a late breakfast? Marvellous organisation, the YHA!'

'And last night?'

'Easy. While you were busy stargazing, I was down by the harbour casting my rod.'

'Well, I've got to hand it to you, you had me fooled the whole time.'

I was beginning to see how his mind worked and why he had acted as he had. I still thought he was stupid, though.

'Why didn't you kick the walk into touch and go off with Dee full time?

'What would you have done?'

'I'd have kicked the walk into touch and gone off with Dee full time.'

'Ha ha. You know what I mean.'

'You really are mad. Why should you give a jam tart what I would have done? I would probably have stayed in Penzance, gone up the Coldstreamer Inn with the two girls you so nicely got rid of for me, poured Moules Marinière over their naughty bits and toasted your good fortune. Please tell me you didn't stick around because you thought I might be scared and lonely without you.'

We set off up the hill, Phil keenly sharing all the lurid details, including that he had told her his surname was McCaverty, and she had actually fallen for it, until he burst out laughing when he was struggling to perform behind the Coverack harbour wall and she really had said 'Come on Phil McCaverty'! I found it hilarious that Phil had been so foxy when it was supposed to be my lucky fox, and it took him a good half an hour to stop chuckling to himself. Even then, when he wasn't wheezing on about the evils of tarmac, he regularly broke into partial repetition of the conversation: 'Bottled it or throttled it, ha ha... are you two an item, hee, hee... Milky Way... a monk... even boats in the harbour go up and down.'

We enjoyed a wonderful flat hilltop for a mile before descending into West, then East Portholland. Here, we met two walkers who had started at Penzance on Saturday, carrying full packs, and were averaging ten miles a day. They were enjoying a pot of tea whilst sitting on the beach wall. How could we resist?

According to the book, and other walkers, the walk round to Mevagissey was supposed to be progressively easier, but we weren't finding it so. A long climb out of East Portholland was followed by ups and downs to Porthluhe Cove, with Caerhays Castle behind. Another long drag up was followed by another, then more ups and downs before the final two ups to Dodman Point from Hemmick Beach. What was happening? Why were we finding it so tough?

We decided a combination of factors were taking effect: we had been walking for nine days in a row; we were carrying full packs; our minds and legs knew this was our last day, and had started to relax, and Phil was just shagged out. Whatever the problem, it was harder than it should have been.

At Dodman Point there was a twenty foot high stone cross with the inscription 'In the firm hope of the second coming of our Lord Jesus Christ, and for the encouragement of those who strive to serve *Him*, this Cross is erected AD 1896'. I decided to make the most of what could be our final opportunity to enjoy the tremendous views we had begun to take for granted.

Inland, china clay pits could be seen shining bright white, a local informing us that one was being converted into the Eden Project, including several biospheres, aimed at education and experimentation. Looking west, we could see as far back as Porthoustock, near Coverack, while in front were our final target, Mevagissey, and far, far away, our original dream, Plymouth – perhaps another time.

The sky and sea were both light grey and still, merging into each other so that boats near the horizon seemed to be floating in the air, with the sun shining dimly through the haze, like a scene from *The Rhyme of the Ancient Mariner*. A gentle zephyr fanned us, completing an idyllic setting.

Phil, ever the romantic, took a final look at the cross, read the inscription aloud, sniffed, and declared that there was not much chance of us making a second coming given how tough we were finding it this time. Off we went again, for the final five miles to Mevagissey.

Round the headland, along Bow Beach, out to Maenease Point and down into the fishing village of Gorran Haven. We had earned a beer, so sought out the pub, which we were informed was at the top of the hill out of town. We moaned, but it wasn't really a close call, and up we went again, to Llawnroc Inn, where food was only served until 2.00p.m. It was now 2.30p.m., so we had a couple of pints sitting in the beer garden watching children play on the grass in the glorious, baking sunshine, before heading back down into town to Cakebread's, a famous local bakers.

Only three miles to go now. Still the lovely path seemed to want to show us some twists in its tail, so up and down we went round Pabyer Point, Turbot Point, Chapel Point, where, at last Portmellon and Mevagissey came into view.

At Portmellon Cove, Phil was delighted to discover the final mile would be on tarmac. Up and over one last headland, then round Stuckumb Point to overlook Mevagissey Harbour. Suddenly, it was over.

We stopped for photographs, mugging a total stranger into recording 'the triumphant intrepid treckers' and my final 'fat photo'. Then we descended into the real world: the bright, fresh, vibrant town of Mevagissey.

It was just gone five, and we had an hour and a half to catch our train at St Austell; which was, coincidentally, the same station we had caught the taxi to Newquay from three weeks earlier. We found our way to the main bus stop, outside 'The Ship Inn', intending to nip in for a celebratory pint once we had checked bus times, but, in a final piece of foxy fortune, a bus marked St Austell came round the corner and pulled up.

I jumped on board.

'Single to St Austell please'.

'Aren't you coming back then?'.

The thought that I might not made me shiver, and left a tingling feeling in my forehead. I turned to share the mood with Phil, who was still standing outside and seemed reluctant to board the bus.

At that very moment, a camper van drove passed, roof rack loaded with all sorts of bits and pieces, including something that vaguely resembled Phil's 'stolen' sleeping bag and mat.

Phil shouted 'Hey, they've got my tent,' and started chasing the van up the road towards Plymouth.

The bus driver gave me a quizzical look, but I motioned him to carry on. As he closed the door and pulled away, I congratulated myself on not getting Phil a ticket, silently wished him good luck in his chase, and settled down for forty winks.

I'd seen the number plate, DEE 6 L, and had a good view of the driver...

6 in Roman numerals is VI...